An Object-Oriented Python Cookbook in Quantum Information Theory and Quantum Computing

This first-of-a-kind textbook provides computational tools in state-of-the-art OOPs Python that are fundamental to quantum information, quantum computing, linear algebra and one-dimensional spin half condensed matter systems. Over 104 subroutines are included, and the codes are aided by mathematical comments to enhance clarity. Suitable for beginner and advanced readers alike, students and researchers will find this textbook to be a helpful guide and a compendium which they can readily use.

Features

- Includes over 104 codes in OOPs Python, all of which can be used either as a standalone program or integrated with any other main program without any issues.
- Every parameter in the input, output and execution has been provided while keeping both beginner and advanced users in mind.
- The output of every program is explained thoroughly with detailed examples.
- Detailed mathematical commenting is done alongside the code which enhances clarity about the flow and working of the code.

Dr. M. S. Ramkarthik works as an assistant professor at the Visvesvaraya National Institute of Technology, Nagpur. His interests are in the fields of theoretical physics, especially in quantum computing, quantum information and mathematical and numerical methods in physics. He likes teaching students at all levels. He is also interested in designing and implementing novel and conceptual methods of teaching physics, making it more palatable and interesting for students. He has also authored the book *Numerical Recipes in Quantum Information Theory and Quantum Computing: An Adventure in FORTRAN 90* published by CRC Press in September 2021. He is a computer enthusiast and interested in solving mathematical and physics problems using various programming languages.

Dr. Pranay Barkataki is currently a Research Analyst in Sony R&D, Bengaluru, where he is working as a data analyst. He has completed his Ph.D. under the supervision of Dr. M. S. Ramkarthik in the field of quantum entanglement and quantum many-body physics. His research interests are in theoretical physics as well as machine learning and AI.

An Object-Oriented Python Cookbook in Quantum Information Theory and Quantum Computing

M. S. Ramkarthik
Pranay Barkataki

CRC Press
Taylor & Francis Group
Boca Raton London New York

CRC Press is an imprint of the
Taylor & Francis Group, an **informa** business

First edition published 2023
by CRC Press
6000 Broken Sound Parkway NW, Suite 300, Boca Raton, FL 33487-2742

and by CRC Press
4 Park Square, Milton Park, Abingdon, Oxon, OX14 4RN

CRC Press is an imprint of Taylor & Francis Group, LLC

Library of Congress Cataloging-in-Publication Data

Names: Ramkarthik, M. S., author. | Barkataki, Pranay, author.
Title: An object oriented Python cookbook in quantum information theory and quantum computing / M.S. Ramkarthik and Pranay Barkataki.
Description: First edition. | Boca Raton : CRC Press, 2023. | Includes bibliographical references and index.
Identifiers: LCCN 2022006256 | ISBN 9781032256078 (hardback) | ISBN 9781032258898 (paperback) | ISBN 9781003285489 (ebook)
Subjects: LCSH: Quantum theory--Mathematics--Data processing. | Quantum computing. | Python (Computer program language) | Object-oriented programming (Computer science)
Classification: LCC QC174.17.M35 R36 2023 | DDC 530.120285/53--dc23/eng20220712
LC record available at https://lccn.loc.gov/2022006256

ISBN: 978-1-032-25607-8 (hbk)
ISBN: 978-1-032-25889-8 (pbk)
ISBN: 978-1-003-28548-9 (ebk)

DOI: 10.1201/9781003285489

Typeset in Latin Modern font
by KnowledgeWorks Global Ltd.

Publisher's note: This book has been prepared from camera-ready copy provided by the authors.

To my dear wife, for her motivation, faith, advice, care and support.

M. S. Ramkarthik

Dedicated to my parents Prabin Kumar Barkataki and Late Runumi Barkataki, who not only taught and nurtured me, but also prepared me to face the challenges in life with strong resolve and humility.

Pranay Barkataki

Contents

Preface

Writing the preface is a ceremonial event for any author. When I wrote the book, *Numerical Recipes in Quantum Information Theory and Quantum Computing: An Adventure in FORTRAN 90*, published by CRC Press in the year 2021, the committee at CRC Press felt that a Python version of this also would be an interesting read and they encouraged me to undertake the project. With this motivation from CRC Press, and also driven by my own "computational" urges, I proceeded to author a book which incorporates a plethora of operations and tasks in quantum information, quantum entanglement and spin chain systems in the latest version of the Python programming language. This book was christened as *An Object-Oriented Python Cookbook for Quantum Information Theory and Quantum Computing*; it is a cookbook because it has a lot of recipes in the aforesaid fields. When I embarked to write this book and develop efficient Python codes, I thought, Why not incorporate the latest data structures in Python like OOPs (Object-Oriented Programming System)? This paved a way to develop codes which not only look clean but also work efficiently.

Python is one of the major tools used by everyone, from software professionals and economists to physicists, chemists, and molecular biologists. The simplicity in syntax and the variety of in-built functions and libraries overshadow its speed compared to other languages. With such superior features and state-of-the-art data structures, OOPs Python was our choice to develop this library. These packages which we have developed can be installed on any computer using a PIP install command without much difficulty. Once it is installed, all the libraries will be in their correct locations (paths) and the users can use the libraries without any problems by following the instructions given in the book. Throughout the book, the SciPy and NumPy libraries are used for invoking certain in-built functions in Python. Another feature of the book is that, for each and every code, we have developed a mathematical way of writing the comments using sophisticated packages so that the user feels at home in every step of the code. This way of developing the comments along the code also has an advantage of establishing a one-to-one mapping between the mathematical nature of the operation and the algorithmic flow in the code as both are quite complementary aspects in solving any problem. At the end of every chapter, a complete example involving a mix of codes from that chapter is illustrated which enhances ones grasp on how to use these codes.

This book contains more than 100 numerical routines in Python and it will be extremely useful for a practicing physicist, both theoretical and experimental, working in the areas of quantum information theory, quantum computing, quantum entanglement and also on condensed matter quantum many-body spin half chains. We also remark here that, there is a chapter devoted to numerical linear algebra procedures and a final chapter on random states and matrices too. We can safely and humbly say that the computational libraries cover a wide spectrum of topics.

In a bird's eye view we see about the chapters. Chapter 1 is a self contained introduction to the Python programming language and the nuances of OOPs programming. This chapter is written in such a way that anyone even without prior knowledge of programming can learn OOPs Python. Each and every syntax and structure of the language is explained with carefully chosen examples for better clarity. Chapter 2 contains the basic tools of quantum mathematics, the mathematical operations which form the core of the theory. Chapter 3 is about numerical linear algebra and other linear algebra–based quantum mechanical operations. Chapter 4 contains those codes with which you can build quantum gates, states and a variety of important quantities involving the states such as fidelity, entropies, trace distance and so on. Chapter 5 deals with quantification of entanglement or entanglement measures. In this chapter, we have developed high-speed methods of computing partial trace and partial transpose of any arbitrary set of qubits from a given superset of qubits. Later, using these operations, we have developed codes for almost all frequently used entanglement measures for both pure and mixed states. Chapter 5 culminates with some illustrations of entanglement detection such as Peres PPT (Positive Partial Transpose) criteria and the Horodecki's reduction criteria for separability. Chapter 6 deals with the construction of several one-dimensional spin half-chain Hamiltonians with and without random bond interactions and with and without homogeneous magnetic fields, including Heisenberg and Ising model and their variants, spin models with asymmetric interactions like the DM interactions. Important to note here is that the codes developed in this chapter will work for any r^{th} neighbour exchange interaction; this enables us to construct any one-dimensional Hamiltonian with very much ease. Last but not least, a final chapter on construction of random states, density matrices and other pseudo random distributions is included for the sake of completeness and elegance.

This book would not have been possible without the help of Ms. Carolina Antunes of CRC Press Taylor & Francis, Ms. Elizabeth (Betsy) Byers of CRC Press and Ms. Sumati Agarwal. All of them were very supportive right from the day the proposal was submitted till the end. I would like to thank CRC Press, Taylor & Francis for giving us this opportunity to write this book and make the codes available to all aspiring researchers in this area. Last but not least, the codes given in this book were verified several times with different physical systems and to the best of our knowledge gives accurate outputs. In spite of several rounds of careful editing and testing, there may always be errors for which the authors will feel grateful if the readers point out the same.

M. S. Ramkarthik
Pranay Barkataki

List of Figures

Acknowledgment

First, I acknowledge my wife Chanchal for providing me a salubrious environment always. I thank my parents for the support they have given me right from my childhood days till now. I give my obeisance to all my teachers, who nurtured me from the day I "consciously" entered into my studies and academics. I thank my friends for the gala time I had spent with them, both during academic and non-academic discussions. At this juncture, I thank the director of my institute for providing me a congenial work atmosphere. My sincere thanks to all my students for being like a grindstone and polishing my ideas constantly. It is my spiritual duty to thank the almighty Lord for being there as my inner strength. This acknowledgement will not be complete if I don't thank the "corona virus" and its "variants" for making me glued to a place, and cutting me off from the external world thereby enabling me to write this book in record time.

M. S. Ramkarthik

I am forever indebted to my parents for giving me opportunities and experiences that have made me who I am today. I would like to specially thank my Ph.D. thesis supervisor Dr. M. S. Ramkarthik for giving me this opportunity to be a co-author in this work. A special thanks to my lab junior Payal Dineshbhai Solanki, for helping me in testing some of the codes. Finally, to all those who have been a part of my getting there: Pranjal Barkataki, Manas Barkataki, Shweta Barkataki, Swagota Barkataki, and Devvrat Tiwari.

Pranay Barkataki

1

Introduction to Python and Object-Oriented Programming

In this chapter we give a standalone introduction to the Object-Oriented Python Programming language. Python name was conceived in the late 1980s by the Dutch programmer Guido van Rossum [1]. Since the first release of Python on February 20, 1991, the Python programming community has grown at a rapid pace, and now a lot of people talk about the eminence of the Python programming language. Therefore, the question arises, what makes Python so much powerful? The answer to this question not only lies in the fact that Python is a general-purpose, dynamic, high-level language, simple and short coding style, but it also behaves like a glue-language, which can interact with other programming languages and their libraries [2], such as C [3], C++ [4] and FORTRAN [5, 6]. These are the reasons which makes Python as the favourite programming language in the field of science [7], engineering [8], data analysis [9], machine learning [10] etc.

However, every strength comes with its weakness, and in this case, it is the execution speed. Python is an interpreted language in which source code does not get directly compiled into machine instructions, rather each statement is executed at run-time by sequentially feeding it into an interpreter loop. In short, the framework of source code implementation in Python is as follows,

1. Translate the source code to an intermediate format known as *byte code* [11]. Byte code provides the advantage of portability, which means that it is a platform-independent format.

2. In the following step, the interpreter interprets the *byte code*.

All the above points convey that Python codes are not compiled directly into binary code like C++ or FORTRAN. Consequently, the codes written in Python are comparatively slower when compared to compiled languages, such as C, which converts the source code to binary code and then executes it. However, counter-arguments to Python being slow are given as follows, rewrite codes with better data structure and algorithms or write performance-intensive parts of the code in any compiled language, and then expose the functionality later in Python. Covering every aspect of Python in this book is impossible, and therefore, in this chapter, we introduce the basic syntax in Python. All codes written in this book run smoothly for Python version 3.7.9 and higher. There are several ways by which we can run a Python program, but the simplest one is through a terminal (or command prompt in Windows). In a terminal, if you type Python at the command prompt, it will display some information about the installed Python version in your computer, for example,

```
$ python
Python 3.7.9 (default, Aug 31 2020, 12:42:55)
[GCC 7.3.0] :: Anaconda, Inc. on linux
Type "help", "copyright", "credits" or "license" for more information.
>>>
```

DOI: 10.1201/9781003285489-1

You can write your Python code just after >>>. Now we will see how to print some character or a number on screen, i.e. on the terminal which we are working on. For example, syntax for printing 'Hello world' on the terminal is given below,

```
>>> print("Hello world")
Hello world
```

If you want to find help at any time, use the help() function. Now to exit back to the command line, you can use the exit() function. A Python code can also be written in a Python script file, and this file has a **.py** file extension. The most common ways to run a Python script file are as follows, from the terminal or integrated development environment (IDE) such as Spyder. To run a Python script from the terminal, you have to open the terminal in the same folder where the Python script which you want to run is located, and thereafter, write the following syntax in the terminal as shown below:

```
$ python <filename>.py
```

For example, if you want to run the Python script *myfile.py* in the terminal then it can be done as follows,

```
$ python myfile.py
```

Usually, most of the Python libraries, including the quantum information library developed in this book, are a collection of functions written in **.py** files. It is now the right time to introduce the structural aspects of any Python code.

1.1   Variables

Variables are reserved memory locations to store values [11]. The values assigned to these variables can be from different data types. For example, it can be number, string, list, tuple, dictionary, etc. Later in this chapter, we will be discussing a few of these data types. A variable name may include the following, uppercase (A-Z) and lowercase (a-z) letters, digits (0-9) and underscore (_). However, a variable name can never start with a number, and we cannot include any of the special characters inside the variable name. Below we have shown few examples of proper variable names,

```
>>> joules = 2    # integer data type
>>> energy_transform = "Convert joules to electron volt"    # string data
    type
>>> Energy_ev = joules*(6.242e+18)    # float data type
>>> probabilities3 = [0.3, 0.6, 0.1]    # List data type
```

Any statement written after # is considered as a comment by the Python interpreter, and this statement is never executed by the interpreter. Comments are written to add more clarity for the person who is trying to decode the program and for a better understanding of the code. Now we are ready to delve into the world of the most commonly used data types in Python.

1.2 Data types

1.2.1 Integer

The *int* class represents integer values. The integer value contains positive or negative whole numbers. In Python, there is no limit to how long an integer value can be. The built-in **type()** function is used to find the data type of any variable, as shown below,

```
>>> a=30
>>> print(type(a))
```

The output of the preceding code will print the *int* class as shown below,

```
<class 'int'>
```

1.2.2 Real

The *float* class represents the real numbers. The real numbers are written in floating point representation. Optionally, the character **e** or **E** followed by a positive or negative integer may be appended to specify scientific notation.

```
>>> b=30.5
>>> print(type(b))
>>> h=6.626E-34 # Planck constant
>>> print("Multiplying Planck constant by 10 =",h*10)
```

The output of the preceding code will print the *float* class as shown below.

```
<class 'float'>
Multiplying Planck constant by 10 = 6.626e-33
```

1.2.3 Complex

Any complex number has two parts, real and imaginary, and it is represented by *complex* class. For example, the complex number $3 + 2i$ in Python can be written as,

```
>>> c = complex(3, 2)
>>> print(type(c))
```

The output of the preceding code will print the *complex* class as shown below,

```
<class 'complex'>
```

1.2.4 String

The *str* class represents the string type. In the Python language, a string is an array of bytes representing Unicode characters. Inside a enclosed single, double or triple quote, we can write any number of characters. Few examples of valid string type variables are shown below,

```
>>> str1 = 'Spooky action at a distance' # Valid string type
>>> str2 = "Entangled 2 qubit states." # Valid string type
```

However, when a string contains a special character like a quote, then we need to escape them, otherwise it will lead to syntax error. Escaping is done by placing backslash (\) before the character as shown below,

```
>>> str3="She said, "Whats up" "   # Not a valid string type
SyntaxError: invalid syntax
>>> str3 ="She said, \"Whats up\" ""   # Valid string type
>>> print(str3)
She said, "Whats up"
>>> str4= 'Space like \t separated'   # \t introduces a tab space
>>> print(str4)
Space like       separated
>>> str5= "Hello \n world"   # \n introduces a new line
>>> print(str5)
Hello
 world
```

One may wonder why in Python the concept of writing a string inside a triple quote is followed. The answer lies in the fact that a triple quote can enclose strings of characters that contain both single and double quotes such that no escaping is needed, and it also allows multi-line strings. Few examples of triple quote strings are shown below,

```
>>> print('''Einstein said, "Quantum mechanics is not correct."''')
Einstein said, "Quantum mechanics is not correct."
>>> print("""Hello
... World""")
Hello
World
```

1.2.5   Logical

There are only two logical values **True** and **False** corresponding to the standard Boolean elements 1 and 0 respectively. The *bool* class represents the Boolean type. An example is indicated as below,

```
>>> print(type(True))
<class 'bool'>
```

1.2.6   List

Lists are an ordered collection of various types of data, and it has no fixed size. Lists need not be homogeneous always, which makes it the most powerful feature of Python. Unlike strings, lists are mutable by a variety of list method calls, and in this section, we will discuss a few of the important ones. A list is created by writing the comma-separated values (items) inside a square bracket, for example,

```
>>> lst1 = [1, 'electron', 2, 'proton', 3, 'neutron']
>>> print(type(lst1))
<class 'list'>
```

In a list, the indexing of the first element starts from zero. If you want to access a particular element or elements in a list, you can do the following:

```
>>> print(lst1[0]) # Indexing by position
1
>>> print(lst1[1:3]) # Slicing a list
['electron', 2]
>>> length = len(lst1) # Calculates the length of the list
>>> print(length)
6
```

To add, remove or insert elements at a particular position in a list can be done by the following list method calls:

```
>>> lst1.append(6) # append element 6 into the list
>>> print(lst1)
[1, 'electron', 2, 'proton', 3, 'neutron', 6]
>>> lst1.remove(3) # remove the element 3
>>> print(lst1)
[1, 'electron', 2, 'proton', 'neutron', 6]
>>> lst1.insert(1,'atom') #.insert(i,x), insert at i'th index the element x
>>> print(lst1)
[1, 'atom', 'electron', 2, 'proton', 'neutron', 6]
>>> lst1.pop(5) # .pop(i), it removes the element in the i'th index
'neutron'
>>> print(lst1)
[1, 'atom', 'electron', 2, 'proton', 6]
```

1.2.7 Tuple

A tuple is a collection of objects which is ordered and immutable. Tuples are sequences, just like lists. The difference between tuples and lists are as follows, tuples cannot be changed like lists, and tuples use parentheses whereas lists use square brackets.

```
>>> tup1 = ('quantum', 'state', 0.75, 0.25) # valid tuple
>>> print(type(tup1))
<class 'tuple'>
>>> tup2 = (1, 2, 3 ) # valid tuple
>>> print(tup2[0])
1
>>> print(tup2[1])
2
```

1.2.8 NumPy arrays

In the heart of any mathematical physics [2] calculation lies an array and, more importantly, in quantum mechanics. For example, arrays represent a quantum state or a density matrix, Hamiltonian of a quantum many-body system, quantum measurement operator, etc. The array data language in Python can be written using the help of NumPy [12], which stands for Numerical Python. It is a library consisting of multidimensional array objects and a collection of routines for processing those arrays. Large parts of NumPy are written in C or C++ for faster computation purposes. NumPy also understands array data format from languages like C and FORTRAN, which is an added advantage. In order to create a NumPy array, we first have to import the library and then call the **array()** method on the sequence of numbers as written below,

```
>>> import numpy as np
>>> array1 = np.array([1, 2, 3]) # 1D array
>>> print(array1.shape) # .shape, it returns the shape of the array
(3,)
>>> array2 = np.array([[1,0],[0,-1]]) # 2D array
>>> print(array2.shape)
(2, 2)
>>> array3 = np.array([[[1,3,2.5],[4,8,-11]],[[3,22,-0.9],[2,3,4]]]) # 3D
    array
>>> print(array3.shape)
(2, 2, 3)
```

There are other ways to create NumPy arrays, and we will discuss a few of them in this section. Primarily we will be discussing about the functions **empty()**, **zeros()**, **identity()** and **ones()**. The **empty()** function takes in an integer or tuple of integer defining the shape of an array and then allocates the memory without assigning any values to the array. It means that random noise close to zero is initialized in the array. The **ones()** and **zeros()** functions take in an integer or tuple of integers defining the shape of an array and return an n-dimensional array whose elements are one or zero, respectively. The **identity()** function takes in an integer value defining the dimension of a square matrix, whose diagonal elements are equal to one and off-diagonal elements are equal to zero. Few examples of the above functions are given below,

```
>>> array3 = np.empty((2,3))
>>> print(type(array3))
<class 'numpy.ndarray'>
>>> array4 = np.zeros((2,3))
>>> print(array4)
[[0. 0. 0.]
 [0. 0. 0.]]
>>> array5 = np.ones((2,3))
>>> print(array5)
[[1. 1. 1.]
 [1. 1. 1.]]
>>> array6 = np.identity(3)
>>> print(array6)
[[1. 0. 0.]
 [0. 1. 0.]
 [0. 0. 1.]]
```

The default data types of outputs of functions **empty()**, **zeros()** and **ones()** is *float64*. The default data type output of **identity()** is *float*. If you want to change the data type from *float* or *float64* to *int* of the above outputs, you can do the following:

```
>>> array7 = np.zeros((2,3), dtype = 'int_') # dtype stands for data type
>>> print(array7)
[[0 0 0]
 [0 0 0]]
>>> array8 = np.ones((2,3), dtype = 'int_')
>>> print(array8)
[[1 1 1]
 [1 1 1]]
>>> array9 = np.identity(3, dtype = 'int_')
>>> print(array9)
[[1 0 0]
 [0 1 0]
 [0 0 1]]
```

There are many varieties of *dtypes* or data types available in NumPy. The *dtypes* have all string character codes and can be used in creating more complicated types. Some *dtypes* are flexible (f), and in these *dtypes* the arrays must have fixed size, but the *dtypes* length may be different for different arrays. Usually string data types are flexible because one array may have strings of length 20 and another array may have strings of length 40.

dtype	Byte	Description
bool_	1	Boolean data type
bool8	1	Alias to bool_
int_		Default integer type. Normally either int64 or int32
int0		Same as int_.
int8	1	Single-byte (8-bit) integer ranging from -128 to 127
byte	1	Alias of int8
int16	2	16-bit integer ranging from -32768 to 32767
int32	4	32-bit integer ranging from -2147483648 to 2147483647
int64	8	64-bit integer ranging from -9223372036854775808 to 9223372036854775807
uint_		Default unsigned integer type; alias to either uint32 or uint64
uint0		Same as uint_
uint8	1	Single-byte (8-bit) unsigned integer ranging from 0 to 255
ubyte	1	Alias of uint8

dtype	Byte	Description
uint16	2	16-bit unsigned integer ranging from 0 to 65535
uint32	4	32-bit unsigned integer ranging from 0 to 4294967295
uint64	8	64-bit unsigned integer ranging from 0 to 18446744073709551615
float_	8	Alias to float64
float16	2	16-bit floating-point number
float32	4	32-bit floating-point number
float64	8	64-bit floating-point number
float96	12	96-bit floating-point number
float128	16	128-bit floating-point number
complex_	16	Alias to complex128
complex64	8	64-bit complex floating-point number
complex128	16	128-bit complex floating-point number
complex256	32	256-bit complex floating-point number
string_	f	Bytes (or str in Python 2) data type. This is a flexible dtype
string0	f	Alias of string_
str_	f	Alias of string_
unicode_	f	String (or Unicode in Python 2) data type. This is a flexible dtype
unicode0	f	Alias of unicode_

1.3   Operators

Operators are objects which perform certain mathematical operations to give another output. Here A and B are called operands. There are three types of operators which can be defined in Python, which are the arithmetic, relational and logical operators as we will see further.

1.3.1 Arithmetic operators

Operator	Description
+	Adds two operands, i.e. $A + B$
−	Subtracts two operands, i.e. $A - B$
*	Multiplies two operands, i.e. $A * B$
/	Performs division, i.e. A/B
**	Raises one operand to the power of the other, i.e. $A * *B = A^B$
%	Performs modulus operation $A\%B$

1.3.2 Relational operators

Operator	Description
==	It checks whether two operands have the same value
! =	It checks whether two operands are unequal
>	It checks whether the left operand is greater than the right operand or not
<	It checks whether the left operand is less than the right operand or not
>=	It checks whether the left operand is greater than or equal to the right operand or not
<=	It checks whether the left operand is less than or equal to the right operand or not

1.3.3 Logical operators

Operator	Description
and	It is called the logical AND operator. If both the operands are non-zero, then the condition becomes true.
or	It is called the logical OR operator. If any one of the two operands are non-zero, then the condition becomes true.
not	It is called the logical NOT operator. Used to reverse the logical state of its operand. If a condition is true then the logical NOT operator will make false and vice versa.

1.4 Decisions

Decisions are very important elements in any programming language. Conditional statements helps us in making those decisions. Basic syntax for conditional statements are given below.

1.4.1 The if statement

The body of the **if** statement starts with an indentation, and the first unindented line marks the end. The generic **if** statement is written as,

```
if <test_expression>:
    <body of if>
```

If the condition <test_expression> holds true then the body of the **if** statement (<body of **if**>) gets executed. An example of the above syntax is given below:

```
# Coefficients of quadratic equation, ax^2 + bx + c = 0
a=5
b=6
c=1
# The discriminant is defined as, b^2 - 4ac
d=(b**2)-(4*a*c)
if d > 0:
    print("Roots are real and distinct")
```

Which will print the output as,

```
Roots are real and distinct
```

1.4.2 The if else statement

The generic **if else** statement is written as,

```
if <test_expression>:
    <body of if>
else:
    <body of else>
end if
```

If the condition <test_expression> holds true then the body of the **if** statement (<body of **if**>) gets executed, otherwise the body of the **else** statement (<body of else>) is executed. An example of the above syntax is given below,

```
# Coefficients of quadratic equation, ax^2 + bx + c = 0
a=5
b=2
c=1
# The discriminant is defined as, b^2 - 4ac
d=(b**2)-(4*a*c)
if d > 0:
```

```
    print("Roots are real and distinct")
else:           # here the condition is d <=0.
    print("Roots are real and equal or both roots are complex")
```

Which will print the output as,

```
Roots are real and equal or both roots are complex
```

1.4.3 The if elif else statement

The abbreviation of the keyword **elif** is *else if*. The generic **if elif else** statement is shown below:

```
if <test_expression0>:
    <Body of if>
elif <test_expression1>:
    <Body of elif 1>
elif <test_expression2>:
    <Body of elif 2>
.
.
.
else:
    <Body of else>
```

The **elif** statements comes after **if** statement and before **else** statement. An example of the above syntax is given below:

```
# Coefficients of quadratic equation, ax² + bx + c = 0
a=5
b=2
c=1
# The discriminant is defined as, b² - 4ac
d=(b**2)-(4*a*c)
if d > 0:
    print("Roots are real and distinct")
elif d == 0:
    print("Roots are real and equal")
else:
    print("Roots are complex")
```

Which will print the output as,

```
Roots are complex
```

1.5 Loops

Loops are useful for repeated execution of similar operations. The importance of looping is so much that, without looping elements, any programming language will be 99% incomplete. There are two types of loops and the syntax of both of them are given below.

1.5.1   The while loop

The generic syntax of the **while** loop is written as,

```
while <test_expression>:
    <Body of while>
```

The body of the **while** loop (<Body of **while**>) gets executed till the condition <test_expression> holds. An example of the above syntax is as follows,

```
# Calculating the n! factorial
n = 3
fact=1
while n > 0:
    fact=fact*n
    n=n-1
print("Factorial =",fact)
```

Which will print the output as,

```
Factorial = 6
```

As we have already discussed above, the **while** loop gets executed till the test expression holds, however, the **while** loop becomes ineffective when you have to iterate over the items of the list or have to run the body of the loop for n times. The for loop fits in this role perfectly as discussed in the next section.

1.5.2   The for loop

The generic syntax of the **for** loop is written as,

```
for <variable> in <iterable>:
    <Body of for loop>
```

The variable <variable> is assigned the value of the new item present in the iterable <iterable> for each loop. Few examples of the above syntax are shown below,

```
mylist=['quantum', 'mechanics', 1, complex(2,3), 3.8]
for i in mylist:
    print(i)
```

Which will print output as,

```
quantum
mechanics
1
(2+3j)
3.8
```

Now if you want to run the variable <variable> from integral number n to integral number m in steps of l (integer number), then you have to use the **range()** function. For the case under consideration, the parametric inputs to the function is as follows, **range(n, m + 1, l)**. Default value of the step in the **range()** function is 1. An example aforesaid case is shown below,

```
# Geometric progression, a, ar, ar², ar³, ar⁴, ..., ar^(n-1)
a=2
r=3
n=5
sum1=0.0
# Loops run from a to ar^(n-1)
for i in range(0,n,1):
    print(a*(r**i))
    sum1=sum1+(a*(r**i))
print("Total sum of geometric progression =",sum1)
```

Which will print output as,

```
2
6
18
54
162
Total sum of geometric progression = 242.0
```

We provide another example below where we show one way to use loops with arrays.

```
# Arithmetic progression, a₁, a₁ + d, a₁ + 2d, ..., a₁ + (n − 1)d
import numpy as np
a1=3
d=0.3
n=5
arr=np.zeros(n,dtype='float64')
for i in range(0,n,1):
    arr[i]=a1+i*d
print("Printing the array holding the arithmetic progression")
print(arr)
```

The above code gives the following output on execution.

```
Printing the array holding the arithmetic progression
[3. 3.3 3.6 3.9 4.2]
```

1.6 Python file handling

This section primarily deals with the inbuilt functions of Python for creating, writing and reading files. There other Python packages such as NumPy [12], PyTables [13], Pandas [14] and Blaze [15], which are not only used for creating, writing and reading files but also performing some advanced operations like cleaning, munging, analysis and visualization. However, in this section, we will restrict the discussion to the inbuilt functions of Python. In Python, two types of files are handled and are text files and binary files. Three file operations take place in the following order when handling a file,

1. Open a file.
2. Read or write (perform the operation).
3. Close the file.

Python has a built-in **open()** function to open a file. This function returns a file object, also called a handle. The generic statement of the **open()** function is shown below,

```
file_object = open(r"file_name", "access_mode")
```

The file should exist in the same directory where the Python program file is saved. Otherwise, before the filename, the full address or path has to be written. The different types of available access modes available in Python are listed below,

Access mode	Code	Description
Read Only	**"r"**	Open text file for reading. The handle is positioned at the beginning of the file. If the file does not exist, then it raises I/O error. This is also the default mode in which file is opened
Read and Write	**"r+"**	Open the file for reading and writing. The handle is positioned at the beginning of the file. Raises I/O error if the file does not exist
Write Only	**"w"**	Open the file for writing. For existing file, the data is truncated and over-written. The handle is positioned at the beginning of the file. Creates the file, if the file does not exists

Access mode	Code	Description
Write and Read	**"w+"**	Open the file for reading and writing. For existing file, data is truncated and over-written. The handle is positioned at the beginning of the file
Append Only	**"a"**	Open the file for writing. The file is created if it does not exist. The handle is positioned at the end of the file. The data being written will be inserted at the end, after the existing data meaning it will be appended
Append and Read	**"a+"**	Open the file for reading and writing. The file is created if it does not exist. The handle is positioned at the end of the file. The data being written will be inserted at the end, after the existing data

An example of the above discussion is shown below,

```
import numpy as np
# To read file MyFile1.txt" in D:\Text in file2
file1 = open(r"D:\Text\MyFile1.txt", "r")
```

```python
# Writing file 'myfile.txt' in the same directory of the python program
file2 = open('myfile.txt', 'w')
L = ["Quantum Information", "\n", "and", "\t", "Computation"]
s = "Numerical recipes\n"
arr=np.array([[0,1],[1,0]])

# Writing a string to file
file2.write(s)

# Writing multiple strings at a time
file2.writelines(L)

# writing matrix elements in the file
file2.write("\nThe matrix is written as\n")
for i in range(0,arr.shape[0]):
    for j in range(0,arr.shape[1]):
        file2.write(str(arr[i,j]))
        # to add tab space length between row elements of matrix
        file2.write('\t')
    # each row is to be printed in the next line
    file2.write('\n')

# Closing file
file2.close()

# Checking if the data is written to file or not
file2 = open('myfile.txt', 'r')
print(file2.read())
file2.close()
```

The file **myfile.txt** stores the following data.

```
Numerical recipes
Quantum Information
and     Computation
The matrix is written as
0       1
1       0
```

The best way to write and load numpy arrays is shown below.

```python
import numpy as np

arr=np.array([[1,2],[3,4],[5,6]])

# Saving the array in a text file
np.savetxt("np_array.txt", arr)

# Displaying the contents of the text file
content = np.loadtxt('np_array.txt')
print("\nContent in np_array.txt:\n", content)
```

```
# Loading the array in different shape
arr2= np.loadtxt('np_array.txt').reshape(2,3)
print("Printing the reshaped matrix")
print(arr2)

# Matrix multiplication of arr and arr2
result=np.matmul(arr,arr2)
print("Multiplication result of two arrays")
print(result)
```

The above code prints the following output.

```
Content in np_array.txt:
[[1. 2.]
 [3. 4.]
 [5. 6.]]
Printing the reshaped matrix
[[1. 2. 3.]
 [4. 5. 6.]]
Multiplication result of two arrays
[[ 9. 12. 15.]
 [19. 26. 33.]
 [29. 40. 51.]]
```

1.7   Math module in Python

The **math** module is a standard module in Python and is always available. To use mathematical functions under this module, you have to import the module using,

```
import math
```

This module does not support complex datatypes. The **cmath** module is the complex counterpart. In the list below all the functions and attributes defined in **math** module and its description are provided

Operator	Description
math.ceil(x)	Returns the smallest integer greater than or equal to x
math.copysign(x, y)	Returns x with the sign of y
math.fabs(x)	Returns the absolute value of x
math.factorial(x)	Returns the factorial of x
math.floor(x)	Returns the largest integer less than or equal to x

Operator	Description
math.fmod(x, y)	Returns the remainder when x is divided by y
math.frexp(x)	Returns the mantissa and exponent of x as the pair (m, e)
math.fsum(iterable)	Returns an accurate floating point sum of values in the iterable
math.isfinite(x)	Returns True if x is neither an infinity nor a NaN (Not a Number)
math.isinf(x)	Returns True if x is a positive or negative infinity
math.isnan(x)	Returns True if x is a NaN.
math.ldexp(x, i)	Returns x * (2**i)
math.modf(x)	Returns the fractional and integer parts of x
math.trunc(x)	Returns the truncated integer value of x
math.exp(x)	Returns e**x
math.expm1(x)	Returns e**x − 1
math.log(x[, b])	Returns the logarithm of x to the base b (defaults to e)
math.log1p(x)	Returns the natural logarithm of 1+x
math.log2(x)	Returns the base-2 logarithm of x
math.log10(x)	Returns the base-10 logarithm of x
math.pow(x, y)	Returns x raised to the power y
math.sqrt(x)	Returns the square root of x
math.cos(x)	Returns the cosine of x
math.sin(x)	Returns the sine of x
math.tan(x)	Returns the tangent of x
math.acos(x)	Returns the arc cosine of x or inverse of cosine of x
math.asin(x)	Returns the arc sine of x or inverse of sine of x
math.atan(x)	Returns the arc tangent of x or inverse of tangent of x
math.cosh(x)	Returns the hyperbolic cosine of x
math.sinh(x)	Returns the hyperbolic sine of x
math.tanh(x)	Returns the hyperbolic tan of x

Operator	Description
math.acosh(x)	Returns the inverse hyperbolic cosine of x
math.asinh(x)	Returns the inverse hyperbolic sine of x
math.atanh(x)	Returns the inverse hyperbolic tan of x
math.atan2(y, x)	Returns atan(y/x)
math.hypot(x, y)	Returns the Euclidean norm, sqrt(x*x + y*y)
math.degrees(x)	Converts angle x from radians to degrees
math.radians(x)	Converts angle x from degrees to radians
math.erf(x)	Returns the error function of x
math.erfc(x)	Returns the complementary error function of x
math.gamma(x)	Returns the Gamma function of x
math.lgamma(x)	Returns the natural logarithm of the absolute value of the Gamma function of x
math.pi	Mathematical constant, the ratio of circumference of a circle to it's diameter (3.14159...)
math.e	Mathematical constant e (2.71828...)

1.8   Functions in Python

If any code is big and spanning several lines, debugging will be very difficult. In these contexts, if we break the program into several parts, the life of the programmer will be relatively easier as the program will look more uncluttered. This work is done by defining the function in the program. Many times it happens that, some part of the program involving a particular operation repeats several times, then, these operations are common to the entire program and it can be defined as a function and called in the main program wherever required. In Python, a function begins with the **def** keyword, thereafter it is followed by the function name, and then within the brackets pass the list of arguments, if any, and then finally a colon (:). The first indented line marks the beginning of the function, and then the first unindented line marks the end of the function. The syntax for the function is given below.

```
def <function_name>(<argument1>, <argument2>, ....):
    <body of the function>
<main body of the program>
```

Note that a function can be non-parameterized, which means that the function takes no input argument. Few examples of functions are shown below,

```python
# This function finds the n!
def factorial(n):
    if n <0:
        return "Cannot find factorial"
    else:
        fact=1
        while n >0:
                fact=fact*n
                n=n-1
        return fact

# Non-parameterized function, and it only prints a statement
def heading():
    print("Finding factorial of a number")

n = 4
# calling the function 'heading'
heading()

# Now we call the function factorial
print('Factorial of the number',n,' =', factorial(n))

# If the number n is negative
n = -1
print('Factorial of the number',n,' =', factorial(n))
```

The output of the previous code is shown below:

```
Finding factorial of a number
Factorial of the number 4  = 24
Factorial of the number -1  = Cannot find factorial
```

1.9   Object-Oriented Programming (OOPs)

A separate chapter can be written on the concept of Object-Oriented Programming (OOPs) [16], however, in this book, the topic is condensed into a section. Many real-world problems can be modelled using OOPs. The codes developed from Chapters 2 to 7 are written in OOPs with state of the art data structure features in Python language. The elemental blocks of OOPs are as follows,

- Object

- Class

- Method

Each elemental block of OOP is explained in detail and compared to real-world examples. The OOP creates an object, and it has two properties as follows, attributes and behaviour. The preceding statement can be understood with the help of an example as follows, Ram,

Matthew, Eric, etc. can be objects, and the attributes of these objects are weight, height, etc., and the behaviour of these objects are as follows, eating, talking, walking, etc. A class is a blueprint for an object, for example, Ram, Matthew and Eric are objects of the class Human. The class defines the common attributes and behaviour of the objects. Methods are the functions defining the behaviour of the objects in a class. Let's try to understand the above discussion by creating the class Human.

1.9.1 Creating the Human class

The object attributes of the Human class are **name, weight** and **height**. The object behaviour are **walking** and **fitness**.

```python
class Human:
    # object attributes
    def __init__(self, name, weight, height):
        self.name = name
        self.weight = weight
        self.height = height

    # object behavior
    def walking(self, hours_walking):
        print(self.name, "walks for", hours_walking, "km")

    # object behavior
    def fitness(self):
        """
        This method calculates whether you are healthy or not.
        Returns:
            fit_level: string type telling whether you are
                        under-weight, fit or over-weight.
        """
        self.BMI = self.weight/(self.height**2)
        if self.BMI < 18.5:
            fit_level = 'under-weight'
        elif self.BMI >= 18.5 and self.BMI <= 24.9:
            fit_level = 'fit'
        else:
            fit_level = 'over-weight'
        return fit_level
```

The **__init__** is a special method, which runs automatically whenever a new object is created. The string written just under the method **fitness** is called a docstring, and it describes the functionality of the method. Using the attribute **__doc__** for any object, we can access the docstring of the method fitness.

1.9.2 Making objects of the Human class

For creating and accessing the attributes of an object in the main body of the program is shown below:

```python
# Creating the objects
person1_obj = Human('Ram', 90, 2.5)
```

```
person2_obj = Human('Matthew', 60, 3)

#Accessing the attributes of the objects
print("Weight of", person1_obj.name, 'is =', person1_obj.weight)
print("Height of", person2_obj.name, 'is =', person1_obj.height)
```

The output of the above code is,

```
Weight of Ram is = 90
Height of Matthew is = 2.5
```

After creating the object, you can access the methods in the class as follows,

```
person1_obj.walking(30)
person2_obj.walking(20)
```

On executing the above code you obtain the following output,

```
Ram walks for 30 km
Matthew walks for 20 km
```

To access the docstring of the method **fitness** you write the following piece of code,

```
# Accessing the docstring
print(person1_obj.fitness.__doc__)
```

The output of the above code is given below:

```
        This method calculates whether you are healthy or not.
        No inputs needed.
        Returns:
            fit_level: string type telling whether you are
                       under-weight, fit or over-weight.
```

A good docstring not only describes the functionality of the method but also describes what type of inputs are expected, and what output the method returns. All the methods written in Chapters 2 to 7 have docstrings for better understanding.

```
print(person1_obj.fitness())
print(person2_obj.fitness())
```

The output of the above code is shown below,

```
fit
under-weight
```

1.9.3 Sphere class

In the previous section, a real-world problem was designed in the OOPs. However, in this section, a more mathematically inclined problem is transformed into the OOPs. To this end, the **Sphere** class is created, and the object attributes are **radius, center_x, center_y** and **center_z**. There are two object behaviours **volume** and **location**. The details of these attributes and behaviours given below:

```python
import math

class Sphere:
    def __init__(self, radius, center_x=0.0, center_y=0.0, center_z=0.0):
        """
        Attributes:
          radius: stores the radius of the sphere
          center_x: x-coordinate of center of sphere, by default = 0.0
          center_y: y-coordinate of center of sphere, by default = 0.0
          center_z: z-coordinate of center of sphere, by default = 0.0
        """
        self.radius=radius
        self.center_x=center_x
        self.center_y=center_y
        self.center_z=center_z

    def volume(self):
        """
        This method calculates the volume of the sphere
        """
        return (4/3)*math.pi*(self.radius**3)

    def location(self,x,y,z):
        """
        This method calculates whether a given point in 3D space
        lies within the sphere or outside.
        Attributes:
          x: the x-coordinate of the  point in a 3D space
          y: the y-coordinate of the  point in a 3D space
          z: the z-coordinate of the  point in a 3D space
        Return: Tells the point lies inside or outside the sphere
        """
        dist = sqrt(((x-self.center_x)**2)+((y-self.center_y)**2)\
            +((z-self.center_z)**2))
        if dist < self.radius:
            return "lies inside the sphere"
        elif dist == self.radius:
            return "lies on the sphere"
        else:
            return "lies outside the sphere"
```

Accessing the class methods are shown below,

```python
# creating the object, where radius of sphere is 1 and centered at (0,0,0)
sphere_obj= Sphere(1)

# Using method volume to calculate the volume of the sphere
vol=sphere_obj.volume()
print("Volume of the sphere is =", vol)

# To find whether point (-2,-1,-3) lies inside sphere or not
```

```
stat= sphere_obj.location(-0.2,-0.1,-0.3)
print("Point (-0.2,-0.1,-0.3)",stat)

# Another object where radius of sphere is 2 and centered at (1,1,1)
sphere_obj2= Sphere(2,1,1,1)

# Using method volume to calculate the volume of the sphere
vol2=sphere_obj2.volume()
print("Volume of the sphere is =", vol2)

# To find whether point (-3,1,2) lies inside sphere or not
stat2= sphere_obj2.location(-3,1,2)
print("Point (-3,1,2)",stat2)
```

The output of the code is shown below:

```
Volume of the sphere is = 4.1887902047863905
Point (-0.2,-0.1,-0.3) lies inside the sphere
Volume of the sphere is = 33.510321638291124
Point (-3,1,2) lies outside the sphere
```

There are primarily four concepts in OOPs, and we briefly introduce these concepts below.

1. **Inheritance**: One class (child class) inherits the attributes and method of another class (parent class).

2. **Encapsulation**: It deals with security, which means that it hides the data access to outsiders.

3. **Polymorphism**: Poly means many and morph means form. It refers to functions or methods having the same name but different functionality.

4. **Data abstraction**: It hides the internal details of a method or function.

From the above discussion, it might seem that OOPs is a powerful style of coding, however, it has its pros and cons. The advantage of using the OOPs style of coding is given below.

- **Improved software-development productivity**: Due to the modularity, extensibility and reusability of OOPs, it is best suited for software-development productivity over traditional procedure-based programming techniques.

- **Faster development**: Due to reusability and a large number of libraries of OOPs which are already available for wider usage makes it easier to develop codes.

- **Improved software maintainability**: It is easier to maintain because of the modular structure of coding. The part of the system which needs to be updated or modified can be achieved without modifying the entire system.

There are a few disadvantages of using the OOPs style of coding, and there are listed below.

- **Large program size**: Compared to procedural programming, the OOPs have extra lines of code.

- **Slower programs**: Compared to procedural programming codes, the OOPs codes require more computation time for execution.

- **Programming complexity**: Due to the inherent complexity of understanding the concepts of OOPs, and therefore many beginner-level programmers do not prefer coding in OOPs style.

1.10   Module and library

A module is a Python file contains different classes or functions. Let us create a module that contains the **Sphere** class from the previous section. The name of the module is **sphere_calc.py**, and save the file in the current directory. Inside the same folder, save and open your Python file and type the following to import the class **Sphere** from the module **sphere_calc.py**,

```
# Importing the class Sphere
from sphere_calc import Sphere

# Another object where radius of sphere is 1 and centered at (1,0,1)
sphere_obj= Sphere(1,1,0,1)

# Using method volume to calculate the volume of the sphere
vol=sphere_obj.volume()
print("Volume of the sphere is =", vol)
```

The above prints the following output,

```
Volume of the sphere is = 4.1887902047863905
```

The collection of many modules make a library or a package. A collection of modules and sub-packages can also make a library. In this book, we construct the **QuantumInformation** library, and it is a collection of 6 modules, and each module is discussed separately from Chapters 2 to 7.

1.11   SciPy

SciPy is an open-source scientific computing library. Many functions in the SciPy library are used in different domains of science and engineering. It contains functions for linear algebra, Ordinary Differential Equations (ODE), integration, Fourier Transform, etc. Primarily we have used the SciPy library for linear algebra applications in our **QuantumInformation** library, and to this end, we will delve more deeply into the module called LAPACK, which primarily contains functions for linear algebra applications.

1.11.1   LAPACK

Calculations involving matrix operations are done by the module called LAPACK (Linear Algebra PACKage) [17, 18]. The library has many functions which are used for solving systems of simultaneous linear equations, least-squares solutions of linear systems of equations, matrix eigenvalue problems, singular value problems, etc. The associated matrix factorizations like (LU, Cholesky, QR, SVD, Schur, generalized Schur) are also provided. The LAPACK module can be imported directly from the SciPy library as shown below,

```
import scipy.linalg.lapack as la
```

Each function has a characteristic name associated with it. In the following table, we provide description of most commonly used functions in LAPACK.

Function	Description
DSYEV	It computes all the eigenvalues and optionally the eigenvectors of a real symmetric matrix
ZHEEV	It computes all the eigenvalues and optionally the eigenvectors of a complex Hermitian matrix
DGEEV	It computes all the eigenvalues and optionally the left or/and right eigenvectors of a real non-symmetric matrix
ZGEEV	It computes all the eigenvalues and optionally left or/and right eigenvectors of a complex non symmetric matrix
DGEQRF	It computes the QR decomposition of a $M \times N$ real matrix
DORGQR	It generates an $M \times N$ real matrix Q with orthonormal columns. **DORGQR** and **DGEQRF** collectively being used for QR decomposition and the output of **DGEQRF** will be fed into **DORGQR**, which will result finally in the orthonormal matrix Q
ZGEQRF	It computes the QR decomposition of a $M \times N$ complex matrix A
ZUNGQR	It generates $M \times N$ complex matrix Q with orthonormal columns. **ZUNGQR** and **ZGEQRF** collectively being used for QR decomposition and the output of **ZGEQRF** will be fed into **ZUNGQR**, which will finally result in the unitary matrix Q
DSYTRD	It reduces a real symmetric matrix A into a tridiagonal matrix T by an orthogonal similarity transformation
DORGTR	It generates a real orthogonal matrix which transforms the given matrix into a tridiagonal form. **DSYTRD** and **DORGTR** collectively being used for tridiagonalization. The output of **DSYTRD** will be fed into **DORGTR** which will finally result in the orthogonal matrix
ZHETRD	It reduces the complex Hermitian matrix A into a real symmetric tridiagonal matrix T by unitary transformation

Function	Description
ZUNGTR	It generates a complex unitary matrix which transforms the given matrix into tridiagonal form. **ZHETRD** and **ZUNGTR** collectively being used for tridiagonalization. The output of **ZHETRD** will be fed into **ZUNGTR**, which will finally result in a unitary matrix
DGETRF	It computes the LU factorization of a $M \times N$ real matrix A using partial pivoting with row interchanges
ZGETRF	It computes the LU factorization of a $M \times N$ general complex matrix A using partial pivoting with row interchanges
ZGETRI	It computes the inverse of a complex matrix using LU factorization as computed by ZGETRF
DGESVD	It computes the singular value decomposition (SVD) of a real $M \times N$ matrix A
ZGESVD	It computes the singular value decomposition (SVD) of a complex $M \times N$ matrix A

More details regarding the input and output parameters of these functions can be found in SciPy library's documentation. For the sake of continuity, few examples on how to use the above functions are shown below,

```python
# importing the LAPACK library
import scipy.linalg.lapack as la

# importing the numpy library
import numpy as np

# real symmetric matrix
A=np.array([[1, 2, 3],[2,4,5],[3,5,6]])

# for diagonalizing the symmetric matrix use DSYEV
eigenvalues, eigenvectors, info= la.dsyev(A)
print("Eigenvalues of symmetric matrix")
print(eigenvalues)
print("Eigenvectors of symmetric matrix")
print(eigenvectors)

# svd of real matrix
U, sigma, VT, info = la.dgesvd(A)

print("The singular values")
print(sigma)
```

The above code prints the following:

```
Eigenvalues of symmetric matrix
```

```
[-0.51572947 0.17091519 11.34481428]
Eigenvectors of symmetric matrix
[[ 0.73697623 0.59100905 0.32798528]
 [ 0.32798528 -0.73697623 0.59100905]
 [-0.59100905 0.32798528 0.73697623]]
The singular values
[11.34481428 0.51572947 0.17091519]
```

1.12   Computation time

In many situations, the user needs to find the computation time for some block of the code or maybe the whole code. The time module in Python is used for calculating the computational time for a block of code.

```python
import time
# Start the time
start=time.time()
# Calculating the computational time of n!
n=50
fact=1
while n >0:
    fact=fact*n
    n=n-1
# Note the time at the end of block of code
end=time.time()
print("Total computational time =",end-start)
```

The computational time depends upon number of system parameters, and therefore the computational time varies from system to system.

1.13   Installation and backends

To install and successfully run the **QuantumInformation** library on your local computer, you have to pre-install the following Python packages,

- NumPy. To install NumPy in your local Linux based system, type the following command in the terminal (command prompt in Windows).

```
pip install numpy
```

- SciPy. To install SciPy in your local Linux based system, type the following command in the terminal

```
pip install scipy
```

The built-in Python packages which we have used in the **QuantumInformation** library are listed below,

- math module.

- cmath module.

- re (Regular Expression) module.

There are primarily two options to install **QuantumInformation** library on your local computer. The first method is through pip install the library, type the following command in terminal (command prompt in Windows) for installing the library on Linux (Windows).

```
pip install QuantumInformation
```

The second method is directly clone the library folder from the following GitHub repository of the co-author, **https://github.com/pranay1990/QuantumInformation.git**. In this method, the folder of the library has to be copied in the same folder as that of the code which is to be compiled.

In order to compile and run a Python program simply open the terminal in the folder where the code is located and type the following command in the terminal.

```
$ python3 <file_name>.py
```

The user can also compile the code using open source integrated development environment such as Spyder, Jupyter Notebook, etc. User can also install the complete open source Python packages and libraries by installing Anaconda from (https://www.anaconda.com/products/individual).

2

Basic Tools of Quantum Mechanics

In the last chapter we have gained an understanding of the Python programming language and its features like OOPs, in this chapter we will proceed with the development of numerical methods for some basic operations of quantum mechanics, which will be helpful in developing other complicated methods (functions). All the methods which will be discussed in this chapter are written inside the class **QuantumMechanics**, and this whole class is written inside the Python module **chap2_nutsboltsquantum.py**. The methods inside the class **QuantumMechanics** also uses some functions from the NumPy and Math libraries. We have assumed that the reader will have a good grasp of basic quantum mechanics [19–23] at the level of post graduation in order to understand the terminologies and the functions in this chapter as well as the entire book without much difficulty. The level of difficulty of functions will increase as we travel through the chapters. Here we have used the standard quantum mechanics convention for all the operations, for example, any pure quantum state is described by a ket vector $| \rangle$ and its corresponding dual by the bra vector which is $\langle |$. In this chapter we have any vector $|v\rangle$ as a $n \times 1$ column matrix, also called "ket v" which can be written explicitly as,

$$|v\rangle = \begin{pmatrix} v_1 \\ v_2 \\ \vdots \\ v_i \\ \vdots \\ v_n \end{pmatrix}.$$

(2.1)

In this book the convention of representing $n \times 1$ pure quantum state or a vector by a column matrix whose shape is equal to $(n,)$. Also note that the Hermitian conjugate of $|v\rangle$ is a $1 \times n$ row matrix called "bra v" written as $\langle v|$, which has an explicit form given by

$$\begin{pmatrix} v_1^* & v_2^* & \cdots & v_i^* & \cdots & v_n^* \end{pmatrix}.$$

(2.2)

In Eq. [2.1] and Eq. [2.2], the entries of the matrix namely the v_i's can either be real or complex corresponding to a real vector and a complex vector respectively. Any one-qubit state $|\psi\rangle$ is written as a linear combination of two bits namely $|0\rangle$ and $|1\rangle$ (eigenstates of the Pauli σ_z matrix) as $|\psi\rangle = \alpha|0\rangle + \beta|1\rangle$ such that $\alpha, \beta \in \mathbb{C}$ and $|\alpha|^2 + |\beta|^2 = 1$. The corresponding dual of $|\psi\rangle$, which is the state $\langle\psi|$ is given by $\alpha^*\langle 0| + \beta^*\langle 1|$. It is well known that one-qubit states are very fundamental in quantum information [24] as multi-qubit states can be built using them. Any physically observable or measurable quantity in

DOI: 10.1201/9781003285489-2

quantum mechanics is represented as a Hermitian operator and thereby it has a matrix representation in a given basis. Keeping these in mind, we have developed methods which involve vector-vector and matrix-matrix operations [25, 26]. Not only that since these multiqubit quantum states can be represented as binary strings, much simplifications can be done in programming by a clever interconversion of binary and decimal entities. To elucidate more, any operation involving a product of $2^N \times 2^N$ matrix with a $2^N \times 1$ column vector, just boils down to an operation of the given operator on the bits rather than the whole 2^N qubit state. The power of such a simplification can be realized when we construct the Hamiltonian matrix.

To import the class **QuantumMechanics** from the **QuantumInformation** library, and creating the object of the class in your Python code can be done as follows,

```
# importing the class QuantumMechanics from the QuantumInformation library
from QuantumInformation import QuantumMechanics as QM

# creating the object of the class
quantum_obj=QM()
```

Regarding the discussion of the functionality of a method, the convention adopted here is – [in] means that the attributes that need to be fed into the method by a user, and [out] means the return of the method to the user. This convention is followed throughout this book.

2.1   Inner product between two vectors

Let there be two n dimensional vectors $|v\rangle$ and $|w\rangle$ in a Hilbert space $\mathcal{H}$ of dimension n defining a physical system, here $|v\rangle$ and $|w\rangle$ are represented by $n \times 1$ column matrices. The inner product between these two vectors can be given in terms of matrix elements as

$$c = \langle v|w\rangle = \sum_{i=1}^{n} v_i^* w_i, \tag{2.3}$$

here, v_i and w_i are the ith entries or matrix elements of vectors $|v\rangle$ and $|w\rangle$ respectively. The inner product is a very important operation in the space $\mathcal{H}$ as it is extensively used to understand properties such as orthogonality and normalization of vectors as these are ubiquitous in quantum mechanics.

2.1.1   For real and complex vectors

```
inner_product(v1,v2)
```

Parameters

In/Out	Argument	Description
[in]	v1	v1 is a NumPy array of dimension $(0:n-1)$
[in]	v2	v2 is a NumPy array of dimension $(0:n-1)$
[out]	inn	inn is complex number storing the inner product between vectors v1 and v2

Implementation

```
def inner_product(self,vec1,vec2):
    """
    Here we compute the inner product
    Attributes:
        vec1: it is a column vector.
        vec2: it is the second column vector.
    Returns:
        inn: inner product between vec1 and vec2.
    """
    inn=complex(0.0,0.0)
    # Checks the dimension of |v> and |w> whether they are equal or not
    assert len(vec1) == len(vec2),\
     "Dimension of two vectors are not equal to each other"

    # Next for loop calculates <v|w>
    for i in range(0,len(vec1)):
        inn=inn+(np.conjugate(vec1[i])*vec2[i])

    return inn
```

Example

In this example, we are finding the inner product between the vectors $v_1 = (1, 2, 3, 4)^T$ and $v_2 = (2, 3, 1, -1)^T$.

```
import numpy as np
v1=np.array([1.0,2.0,3.0,4.0])
v2=np.array([2.0,3.0,1.0,-1.0])
inn=quantum_obj.inner_product(v1,v2)
print(inn)
```

Prints, to standard output which is the inner product between the two real vectors.

```
(7+0j)
```

In this example, we are finding the inner product between the vectors $v_1 = (1, 1 - i, 2 + 3i)^T$ and $v_2 = (2, 3 - i, 2 - 3i)^T$.

```
import numpy as np
```

```
v1=np.array([1, complex(1,-1), complex(2,3)])
v2=np.array([2, complex(3,-1), complex(2,-3)])
inn=quantum_obj.inner_product(v1,v2)
print(inn)
```

Prints, to standard output which is the inner product between the two complex vectors.

```
(1-10j)
```

2.2   Norm of a vector

For a given n dimensional vector $|v\rangle$ in the Hilbert space, using the inner product of the vectors as given by Eq. [2.3], the Euclidean norm of the vector $|v\rangle$ can be defined as

$$||v|| = \sqrt{\langle v|v\rangle} = \sqrt{\sum_{i=1}^{n} v_i^* v_i} = \sqrt{\sum_{i=1}^{n} |v_i|^2}, \qquad (2.4)$$

the norm is a very important number in the Hilbert space, it is used as a measure to define the length of the abstract vector $|v\rangle$.

2.2.1   For a real or complex vector

```
norm_vec(vec)
```

Parameters

in/out	Argument	Description
[in]	vec	vec is a NumPy array of dimension (0:n−1)
[out]	norm	norm is the norm of the vector named vec

Implementation

```
def norm_vec(self,vec):
    """
    Here we calculate norm of a vector
    Attributes:
        vec: column vector
    Returns:
        norm: it contains the norm of the column vector vec
    """
    norm=0.0
    # Next for loop calculates the ||v|| = √⟨v|v⟩
    for i in range(0,len(vec)):
```

```
        norm=norm+abs(np.conjugate(vec[i])*vec[i])
    return np.sqrt(norm)
```

Example

In this example, we are finding the norm of the complex vector $v = (1, 2, 3, 4)^T$

```
import numpy as np
v=np.array([1.0,2.0,3.0,4.0])
norm=quantum_obj.norm_vec(v)
print(norm)
```

Prints, to standard output which is the norm of the vector.

```
5.477225575051661
```

In this example, we are finding the norm of the vector $v = (1, 1 - i, 2 + 3i)^T$.

```
import numpy as np
v=np.array([1,complex(1,-1),complex(2,3)])
norm=quantum_obj.norm_vec(v)
print(norm)
```

Prints, to standard output which is the norm of the vector.

```
4.0
```

2.3   Normalization of a vector

Once the norm is found using Eq. [2.4], we can proceed to normalize the n dimensional vector $|v\rangle$ to obtain the normalized vector $|v'\rangle$ as

$$|v'\rangle = \frac{|v\rangle}{||v||}. \tag{2.5}$$

Normalization is an important operation which is done on quantum states so as to make the length of the vector representing the state to have unit length, that is to say, even though $\langle v|v \rangle$ may not be unity in general, it is guaranteed that $\langle v'|v' \rangle$ is 1.

2.3.1   For a real or complex vector

```
def normalization_vec(self,vec)
```

Parameters

in/out	Argument	Description
[in]	vec	vec is a NumPy array of dimension (0:n−1)
[out]	vec	vec is a normalized NumPy array of dimension (0:n−1)

Implementation

```
def normalization_vec(self,vec):
    """
    Here normalize a given vector
    Attributes:
        vec: unnormalized column vector
    Returns:
        vec: normalized column vector
    """
    norm=0.0
    # Next for loop calculates ||v||
    for i in range(0,len(vec)):
        norm=norm+abs(np.conjugate(vec[i])*vec[i])
    # Next calculates |v⟩/||v||
    vec=vec/np.sqrt(norm)
    return vec
```

Example

In this example, we are normalizing the real vector $v = (1, 2, 3, 4)^T$.

```
vec=np.array([1,2,3,4])
vec=quantum_obj.normalization_vec(vec)
print(vec)
```

Prints, to standard output which is the normalized vector.

```
[0.18257419 0.36514837 0.54772256 0.73029674]
```

In this example, we are normalizing the complex vector $v = (1, 1 - i, 2 + 3i)^T$.

```
vec = np.array([1,complex(1,-1),complex(2,3)])
vec=quantum_obj.normalization_vec(vec)
print(vec)
```

Prints, to standard output which is the normalized vector.

```
[0.25+0.j 0.25-0.25j 0.5 +0.75j]
```

2.4   Outer product of a vector

If we have two vectors $|v\rangle$ and $|w\rangle$ of dimensions n_1 and n_2 respectively, the outer product between them is an operator which has the matrix representation A having the dimension $n_1 \times n_2$ given by,

$$A = |v\rangle\langle w|, \tag{2.6}$$

and the ij'th matrix element of A is given by,

$$A_{ij} = \langle i|A|j\rangle = v_i w_j^*. \tag{2.7}$$

The outer product is an important entity in quantum mechanics which has many uses, specially used in the representation of projection operators, density matrices and so on. When $|v\rangle = |w\rangle$, it represents the projection operator.

2.4.1 For real vectors

`outer_product_rvec(vec1,vec2)`

Parameters

in/out	Argument	Description
[in]	vec1	vec1 is a real NumPy array of dimension (0:n1−1)
[in]	vec2	vec2 is a real NumPy array of dimension (0:n2−1)
[out]	matrix	matrix is a real NumPy array of dimension (0:n1−1,0:n2−1) and it stores the outer product

Implementation

```python
def outer_product_rvec(self,vec1,vec2):
    """
    Here we calculate the outer product
    Attributes:
        vec1: it is a column real vector
        vec2: it is another real vector
    Returns:
        matrix: outer product of vec1 and vec2
    """
    matrix = np.zeros([len(vec1),len(vec2)],dtype='float64')
    # In the next for loop we calculate |v⟩⟨w|
    for i in range(0,len(vec1)):
        for j in range(0,len(vec2)):
            matrix[i,j]=vec1[i]*vec2[j]
    return matrix
```

Example

In this example, we are finding the outer product between real vectors $v_1 = (1,2)^T$ and $v_2 = (2,3,1)^T$.

```python
import numpy as np
vec1=np.array([1,2])
vec2=np.array([2,3,1])
matrix=quantum_obj.outer_product_rvec(vec1,vec2)
print(matrix)
```

Prints, to standard output which is the outer product of vectors vec1 and vec2.

```
[[2. 3. 1.]
 [4. 6. 2.]]
```

2.4.2   For complex vectors

`outer_product_cvec(vec1,vec2)`

Parameters

in/out	Argument	Description
[in]	vec1	vec1 is a complex NumPy array of dimension $(0{:}n1-1)$
[in]	vec2	vec2 is a complex NumPy array of dimension $(0{:}n2-1)$
[out]	matrix	matrix is a complex NumPy array of dimension $(0{:}n1-1,0{:}n2-1)$ and it stores the outer product

Implementation

```python
def outer_product_cvec(self,vec1,vec2):
    """
    Here we calculate the outer product
    Attributes:
        vec1: it is a column complex vector
        vec2: it is another complex vector
    Returns:
        matrix: outer product of vec1 and vec2
    """
    matrix = np.zeros([len(vec1),len(vec2)],dtype=np.complex_)
    # Next for loop we calculate |v><w|
    for i in range(0,len(vec1)):
        for j in range(0,len(vec2)):
            matrix[i,j]=vec1[i]*np.conjugate(vec2[j])
    return matrix
```

Example

In this example, we are finding the outer product between vectors $v_1 = (1, 1-i)^T$ and $v_2 = (2, 3-i)^T$.

```python
import numpy as np
vec1=np.array([1,complex(1,-1)])
vec2=np.array([2,complex(3,-1)])
matrix=quantum_obj.outer_product_cvec(vec1,vec2)
print(matrix)
```

The above code prints the following standard output of outer product,

```
[[2.+0.j 3.+1.j]
 [2.-2.j 4.-2.j]]
```

2.5 Tensor product between two matrices

Tensor product is also known as direct product or Kronecker product. Tensor products are not restricted to square matrices only, but valid even for rectangular matrices equally. Let there be two matrices A and B of the dimensions $m \times n$ and $p \times q$ respectively. The Kronecker product between these two matrices will be a matrix of dimension $mp \times nq$ can be represented as below,

$$
A \otimes B = \begin{pmatrix} A_{11}B & A_{12}B & \cdots & A_{1n}B \\ A_{21}B & A_{22}B & \cdots & A_{2n}B \\ \vdots & \vdots & \ddots & \vdots \\ A_{m1}B & A_{m2}B & \cdots & A_{mn}B \end{pmatrix}_{mp \times nq} . \tag{2.8}
$$

Note that each of the elements of the matrix A is algebraically multiplied by the total matrix B. Kronecker products are very useful when we deal with several identical or non identical physical systems, and to find the quantum states representing those systems. For example, any general one-qubit system can be represented by the state $|\phi\rangle = \alpha|0\rangle + \beta|1\rangle$, now, if we have to find the two-qubit state $|\xi\rangle$ corresponding to two one-qubit states, then we can have

$$
|\xi\rangle = (\alpha_1|0\rangle + \beta_1|1\rangle)_1 \otimes (\alpha_1|0\rangle + \beta_1|1\rangle)_2. \tag{2.9}
$$

The subscripts in the above refer to systems 1 and 2 respectively. Note that Eq.[2.9] can be extended to N qubit states in a straightforward manner. Not only that, the concept of tensor product can be extended to operators, which represent physical observables in quantum mechanics also.

2.5.1 For real or complex matrices

```
tensor_product_matrix(A,B)
```

Parameters

in/out	Argument	Description
[in]	A	A is a NumPy 1D or 2D array
[in]	B	B is a NumPy 1D or 2D array
[out]	C	C is a NumPy array which is the tensor product of matrix A and B

Implementation

```
def tensor_product_matrix(self,A,B):
    """

    Here we calculate the tensor product of two matrix A and B
    Attributes:
        A: it is either a 1D or 2D array
        B: it is either a 1D or 2D array
    Returns: tensor product of A and B
    """
    # Reshaping if input vector is of shape $(n,)$ to $(n,1)$.
    if len(A.shape)==1:
        A=A.reshape(A.shape[0],1)
    if len(B.shape)==1:
        B=B.reshape(B.shape[0],1)
    return np.kron(A,B)
```

Example

In this example, we are finding the tensor product between two real matrices as given below:

$$A = \begin{pmatrix} 1 \\ 2 \end{pmatrix}, \quad B = \begin{pmatrix} 1 & 2 \\ 3 & 1 \end{pmatrix} \tag{2.10}$$

```
import numpy as np
A=np.array([1,2])
B=np.array([[1,2],[3,1]])
C=quantum_obj.tensor_product_matrix(A,B)
print(C)
```

The above code prints the output of the tensor product between the matrix A and B.

```
[[1 2]
 [3 1]
 [2 4]
 [6 2]]
```

In this example, we are finding the tensor product between two complex matrices as given below:

$$A = \begin{pmatrix} 1 \\ 1 - i \end{pmatrix}, \quad B = \begin{pmatrix} 1 + i \\ 3 - 3i \end{pmatrix} \tag{2.11}$$

```
import numpy as np
A=np.array([1,complex(1,-1)])
B=np.array([complex(1,1),complex(3,-3)])
C=quantum_obj.tensor_product_matrix(A,B)
print(C)
```

The above code prints the output of the tensor product between the matrix A and B.

```
[[1.+1.j]
 [3.-3.j]
 [2.+0.j]
 [0.-6.j]]
```

2.6   Commutator between two matrices

In general it is known that matrices do not commute under multiplication, so it is useful to define the quantity called commutator represented by $[A, B]$ between two matrices and is given by,

$$C = [A, B] = AB - BA, \tag{2.12}$$

if $[A, B] = 0 \Rightarrow AB = BA$, then matrix A and B are said to commute with each other. Commutator arises as a basic principle in quantum mechanics where we start with the cannonical commutation relation [27] between the i'th and j'th component of position and momentum respectively given by $[x_i, p_j] = i\hbar\delta_{ij}$ from which the entire edifice of quantum mechanics is built. It is also worth to mention here that if two physical observables do not commute, then they cannot be diagonalized by a common set of eigenvectors, this has deep implications to Heisenberg uncertainty principle etc.

2.6.1   For real or complex matrices

commutation(A,B)

Parameters

in/out	Argument	Description
[in]	A	A is a NumPy array of dimension (0:n−1,0:n−1)
[in]	B	B is a NumPy array of dimension (0:n−1,0:n-1)
[out]	C	C is a NumPy array of dimension (0:n−1,0:n−1) which is the commutator between A and B

Implementation

```
def commutation(self,A,B):
    """
    Here we calculate commutation between matrices A and B
    Attributes:
        A: it is a matrix
        B: it is another matrix
    Returns: AB-BA matrix
    """
```

```
We directly return the commutation relation, AB − BA
return np.matmul(A,B)-np.matmul(B,A)
```

Example

In this example, we are finding the commutator between two real matrices as given below:

$$A = \begin{pmatrix} 1 & 2 \\ 2 & -1 \end{pmatrix}, \quad B = \begin{pmatrix} 1 & 2 \\ 3 & 1 \end{pmatrix} \tag{2.13}$$

```
import numpy as np
A=np.array([[1,2],[2,-1]])
B=np.array([[1,2],[3,1]])
C=quantum_obj.commutation(A,B)
print(C)
```

The above code prints the output of the commutator between matrices A and B.

```
[[ 2  4]
 [-6 -2]]
```

In this example, we are finding the commutator between two complex matrices as given below,

$$A = \begin{pmatrix} 1 & 1-i \\ 2+3i & 4 \end{pmatrix}, \quad B = \begin{pmatrix} 1+i & 1+i \\ 3-3i & 4-i \end{pmatrix} \tag{2.14}$$

```
import numpy as np
A=np.array([[1,complex(1,-1)],[complex(2,3),4]])
B=np.array([[complex(1,1),complex(1,1)],[complex(3,-3),complex(4,-1)]])
C=quantum_obj.commutation(A,B)
print(C)
```

The above code prints the output of the commutator between matrices A and B.

```
[[ 1.-11.j -2. -8.j]
 [-3.-14.j -1.+11.j]]
```

2.7   Anticommutator between two matrices

The anticommutator $\{A, B\}$ between two matrices A and B is defined as below,

$$C = \{A, B\} = AB + BA, \tag{2.15}$$

if $\{A, B\} = 0 \Rightarrow AB = -BA$, then matrix A and B are said to anti-commute with each other. Anticommutators occur in the areas of second quantization involving Fermionic and Bosonic operators [28].

2.7.1   For real or complex matrices

`anti_commutation(A,B)`

Parameters

in/out	Argument	Description
[in]	A	A is a NumPy array of dimension $(0{:}n{-}1,0{:}n{-}1)$
[in]	B	B is a NumPy array of dimension $(0{:}n{-}1,0{:}n{-}1)$
[out]	C	C is a NumPy array of dimension $(0{:}n{-}1,0{:}n{-}1)$ which is the anticommutator between A and B

Implementation

```
def anti_commutation(self,A,B):
    """

    Here we calculate anti commutation between matrices A and B
    Attributes:
        A: input a square matrix
        B: input another square matrix
    Returns: AB+BA matrix
    """
    # The following return statement calculates, {A, B} = AB + BA
    return np.matmul(A,B)+np.matmul(B,A)
```

Example

In this example, we are finding the anticommutator between two real matrices as given below:

$$A = \begin{pmatrix} 1 & 2 \\ 2 & -1 \end{pmatrix}, \quad B = \begin{pmatrix} 1 & 2 \\ 3 & 1 \end{pmatrix} \tag{2.16}$$

```
import numpy as np
A=np.array([[1,2],[2,-1]])
B=np.array([[1,2],[3,1]])
C=quantum_obj.anti_commutation(A,B)
print(C)
```

The above code prints the output of the anticommutator between matrices A and B.

```
[[12  4]
 [ 4  8]]
```

In this example, we are finding the anticommutator between two complex matrices as given below:

$$A = \begin{pmatrix} 1 & 1-i \\ 2+3i & 4 \end{pmatrix}, \quad B = \begin{pmatrix} 1+i & 1+i \\ 3-3i & 4-i \end{pmatrix} \tag{2.17}$$

```
import numpy as np
A=np.array([[1,complex(1,-1)],[complex(2,3),4]])
B=np.array([[complex(1,1),complex(1,1)],[complex(3,-3),complex(4,-1)]])
C=quantum_obj.anti_commutation(A,B)
print(C)
```

The above code prints the output of the anticommutator between matrices A and B.

```
[[ 1.+1.j 10.+0.j]
 [25.+0.j 31.-9.j]]
```

2.8　Binary to decimal conversion

In mathematics, binary numbers are expressed in the *base-2* number system, which only uses '0' and '1' to represent them, which are called bits. Any binary number can be easily converted to a decimal using bit values as follows. Let us have a binary string $[a_1 a_2 a_3 \cdots a_n]_2$ where values a_i's can be either 0 or 1 only. To convert a binary string into the decimal equivalent we have,

$$[a_1 a_2 a_3 \cdots a_n]_2 = a_1 2^{n-1} + a_2 2^{n-2} + \cdots + a_{n-1} 2^1 + a_n 2^0, \qquad (2.18)$$

here the right to left convention is followed. Binary to decimal conversion is a very important operation in quantum mechanics and quantum information theory because, the states of the quantum systems (meaning pure states) are represented by a string of zeros and ones. For example, the state of two electrons can be represented as $|00\rangle = |\uparrow\uparrow\rangle$, $|01\rangle = |\uparrow\downarrow\rangle$, $|10\rangle = |\downarrow\uparrow\rangle$ and $|11\rangle = |\downarrow\downarrow\rangle$. When we have to manipulate such states, then the bit representation of such states and its corresponding decimal equivalent becomes handy. For example, the decimal equivalent of the above states can be written as $|00\rangle = |0\rangle$, $|01\rangle = |1\rangle$, $|10\rangle = |2\rangle$ and $|11\rangle = |3\rangle$. Such representations can be easily extended to multi-qubit states, for example, a state with four electrons can be in one of the states given by $|1010\rangle = |\downarrow\uparrow\downarrow\uparrow\rangle = |10\rangle$. The main advantage of this decimal representation is that, instead of dealing with a string of numbers which may often be tiring, we deal only with one number.

```
binary_decimal(vec)
```

Parameters

in/out	Argument	Description
[in]	vec	vec is a NumPy array containing binary matrix elements i.e. 0 or 1
[out]	dec	dec is decimal equivalent of the NumPy array vec

Implementation

```
def binary_decimal(self,vec):
    """
    Binary to decimal conversion
    Input:
        vec: array containing binary matrix elements i.e. 0 or 1
    Returns:
        dec: decimal equivalent number of binary array vec
    """
    dec=0
    for i in range(0,vec.shape[0]):
        dec=dec+int((2**(vec.shape[0]-1-i))*vec[i])
    dec = int(dec)
    return dec
```

Example

In this example, we are converting binary string of numbers 1010 into a decimal number.

```
A=np.array([1,0,1,0])
B=quantum_obj.binary_decimal(A)
print(B)
```

The above code prints the output of the decimal equivalent of 1010.

```
10
```

2.9   Decimal to binary conversion

To convert an integer number n into a binary string, we adopt the following algorithm,

1. Divide the integer n by 2.

2. Store the remainder from step 1 as the least significant number of the binary string.

3. Divide the quotient (which we got from previous step) by 2, assign the remainder as the second least significant number.

4. Repeat the process till we get the quotient to be zero.

The final remainder will be the most significant digit of the binary string.

```
decimal_binary(i,N)
```

Parameters

in/out	Argument	Description
[in]	i	decimal number for which binary equivalent to be calculated
[in]	N	number of the binary string
[out]	vec	binary equivalent of the decimal number i

Implementation

```
def decimal_binary(self,i,N):
    """
    Decimal to binary conversion
    Inputs:
        i: decimal number for which binary equivalent to be calculated
        N: length of the binary string
    Returns:
        bnum: binary number equivalent to i, it is column matrix.
    """
    bnum=np.zeros([N],dtype=int)
    for j in range(0,N):
        bnum[bnum.shape[0]-1-j]=i%2
        i=int(i/2)
    return bnum
```

Example

In this example we are converting the integer 10 into a binary string number.

```
dec=10
# 4 string
vec=quantum_obj.decimal_binary(dec,4)
print("The four binary string of 10")
print(vec)
# 6 string
vec2=quantum_obj.decimal_binary(dec,6)
print("The six binary string of 10")
print(vec2)
```

The above code prints the output of the binary equivalent of 10.

```
The four binary string of 10
[1 0 1 0]
The six binary string of 10
[0 0 1 0 1 0]
```

2.10   Binary digits shift

This algorithm is used in shifting the string of binary digits to the left or right by k place values. Let us define the left and right shift operators as, S_L and S_R. The operation of the left shift operator is shown below,

$$S_L[a_1 a_2 a_3 \cdots a_n]_2 = [a_2 a_3 a_4 \cdots a_n a_1]_2. \tag{2.19}$$

The operation of the right shift operator is shown below,

$$S_R[a_1 a_2 a_3 \cdots a_n]_2 = [a_n a_1 a_2 a_3 a_4 \cdots a_{n-1}]_2. \tag{2.20}$$

In Eq. [2.19] and Eq. [2.20], the a_i's takes value either zero or one in the binary representation. For shifting left or right k times we have to operate S_L or S_R, respectively, totally k times in succession. This routine is useful for studying quantum states which are shift symmetric and it can be put to other generic uses depending on the problem under consideration.

`binary_shift(vec,shift=1,shift_direction='right')`

Parameters

in/out	Argument	Description
[in]	vec	vec is an integer array of dimension (0:n−1) which is the given binary string
[in]	shift	It is degree of the shift towards left or right direction. By default the value is 1
[in]	shift_direction	It can take only the following values, • shift_direction = 'right' $\Rightarrow$ right direction shift, and it is the default value. • shift_direction = 'left' $\Rightarrow$ left direction shift.
[out]	shift_vec	It is an integer array of dimension (0:n−1) which is the shifted binary string of the array vec

Implementation

```
def binary_shift(self,vec,shift=1,shift_direction='right'):
    """
    Shifting of string of binary number
    Input:
        vec: array containing binary matrix elements i.e. 0 or 1
        shift: degree of the shift
        shift_direction: its value is either left or right
    Output:
```

```
        shift_vec: Shifted vec
    """
    # Following statement check whether direction entered is left or right.
    assert shift_direction == 'left' or shift_direction == 'right',\
    'Not proper shift direction'
    # Following statement check that degree of shift is with [1,n-1]
    assert shift >= 1 and shift < vec.shape[0],\
    "degree of shift is not proper"
    # Following array will store the shifted binary number
    shift_vec=np.zeros([vec.shape[0]],dtype='int_')
    if shift_direction =='right':
        # for loop calculates the right direction shift
        for i in range(0,vec.shape[0]):
            ishift=i+shift
            if ishift >= vec.shape[0]:
                ishift=ishift-vec.shape[0]
            shift_vec[ishift]=vec[i]
    if shift_direction == 'left':
        # for loop calculates the left direction shift
        for i in range(0,vec.shape[0]):
            ishift=i-shift
            if ishift < 0:
                ishift=ishift+vec.shape[0]
            shift_vec[ishift]=vec[i]
    return shift_vec
```

Example

In this example, we shift the binary string 0110 to the left by two degrees, and print the output. Next, we shift the binary string 0110 to the right by one degree, and print the output.

```
import numpy as np
A=np.array([0,1,1,0,0])
B=quantum_obj.binary_shift(A,2,'left')
C=quantum_obj.binary_shift(A,1,'right')
print("Left direction shift")
print(B)
print("Right direction shift")
print(C)
```

The above code prints the following output for both the above cases.

```
Left direction shift
[1 0 0 0 1]
Right direction shift
[0 0 1 1 0]
```

2.11 Quantum state shift

In this section, we develop a numerical function to evaluate the translational shift of a quantum state, spanned by the computational basis. The idea behind the translational shift of a quantum state is explained with the help of an example, consider a three-qubit quantum state which is written as follows,

$$|\psi\rangle = c_1|001\rangle + c_2|010\rangle + c_3|100\rangle, \tag{2.21}$$

where $|c_1|^2 + |c_2|^2 + |c_3|^2 = 1$. The matrix representation of the above quantum state in the computational basis can be written as a 8×1 column vector as shown below.

$$|\psi\rangle = \begin{pmatrix} 0 \\ c_1 \\ c_2 \\ 0 \\ c_3 \\ 0 \\ 0 \\ 0 \end{pmatrix}. \tag{2.22}$$

The left and right translation operators are defined as follows, $\hat{T}_L$ and $\hat{T}_R$, respectively. The result on operating $\hat{T}_L$ on the quantum state written in Eq. [2.21] is shown below,

$$\hat{T}_L|\psi\rangle = |\psi_L\rangle = c_1|010\rangle + c_2|100\rangle + c_3|001\rangle. \tag{2.23}$$

Similarly, on operating $\hat{T}_R$ on the quantum state written in Eq. [2.21], we obtain the following,

$$\hat{T}_R|\psi\rangle = |\psi_R\rangle = c_1|100\rangle + c_2|001\rangle + c_3|010\rangle. \tag{2.24}$$

The matrix representation of the states $|\psi_L\rangle$ and $|\psi_R\rangle$ are shown below,

$$|\psi_L\rangle = \begin{pmatrix} 0 \\ c_3 \\ c_1 \\ 0 \\ c_2 \\ 0 \\ 0 \\ 0 \end{pmatrix}, \qquad |\psi_R\rangle = \begin{pmatrix} 0 \\ c_2 \\ c_3 \\ 0 \\ c_1 \\ 0 \\ 0 \\ 0 \end{pmatrix}. \tag{2.25}$$

For translating the quantum state left or right k times, we have to operate $\hat{T}_L$ or $\hat{T}_R$, respectively, totally k times in succession.

2.11.1   For a real state

---

rstate_shift(vec,shift=1,shift_direction='right')

Parameters

in/out	Argument	Description
[in]	vec	vec is a real quantum state
[in]	shift	It is the degree of the translational shift towards left or right direction. By default, the value is 1
[in]	shift_direction	It can take only the following values, • shift_direction = 'right' ⇒ right direction shift, and it is the default value • shift_direction = 'left' ⇒ left direction shift
[out]	state2	It is the translationally shifted quantum state of the state vec

Implementation

```
def rstate_shift(self,vec,shift=1,shift_direction='right'):
    """
    Shifting of a real quantum state
    Inputs:
        vec: real quantum state.
        shift: degree of the state.
        shift_direction: It shows the direction of the shift,
                    by default it is right
    Return:
            state2: shifted state of vec
    """
    # Checks whether entered state is written in 2^N vector space
    assert vec.shape[0]%2==0,"Not a qubit quantum state"
    # N stores the number of qubits
    N=int(math.log(vec.shape[0],2))
    # basis vector stores binary strings of any computational basis
    basis=np.zeros([N],dtype=int)
    # state2 will store the shifted quantum state
    state2=np.zeros([2**N],dtype='float64')
    # Next for loop calculates, $\hat{T}_L|\psi\rangle$ or $\hat{T}_R|\psi\rangle$, as per choice
    for i in range(0,vec.shape[0]):
        if vec[i] != 0.0:
            basis=self.decimal_binary(i,N)
            basis=self.binary_shift(basis,shift=shift,\
                            shift_direction=shift_direction)
```

```
        j=int(self.binary_decimal(basis))
        state2[j]=vec[i]

    return state2
```

Example

In this example, we left shift the following quantum state by 1 position to left

$$|\psi\rangle = 0.218218|001\rangle + 0.436436|010\rangle + 0.872872|100\rangle. \qquad (2.26)$$

```
A = np.array([0, 0.218218, 0.436436, 0, 0.872872, 0, 0, 0])
A=quantum_obj.normalization_vec(A)
B=quantum_obj.rstate_shift(A,shift=1,shift_direction='left')
print(B)
```

The above code prints the left shifted quantum state as shown below:

```
[0. 0.87287156 0.21821789 0. 0.43643578 0. 0. 0.]
```

2.11.2 For a complex state

```
cstate_shift(vec,shift=1,shift_direction='right')
```

Parameters

in/out	Argument	Description
[in]	vec	vec is a complex quantum state
[in]	shift	It is the degree of translational shift towards left or right direction. By default, the value is 1
[in]	shift_direction	It can take only the following values, • shift_direction = 'right' ⇒ right direction shift, and it is the default value • shift_direction = 'left' ⇒ left direction shift
[out]	state2	It is the translationally shifted quantum state of the state vec

Implementation

```
def cstate_shift(self,vec,shift=1,shift_direction='right'):
    """
    Shifting of a complex quantum state
    Inputs:
```

```
        vec: complex quantum state.
        shift: degree of the state.
        shift_direction: It shows the direction of the shift, by default
                         it is right
    Return:
        state2: shifted state of vec
    """
```

```python
# Checks whether entered state is written in 2^N vector space
assert vec.shape[0]%2==0,"Not a qubit quantum state"
# N stores the number of qubits
N=int(math.log(vec.shape[0],2))
# basis vector stores binary strings of any computational basis
basis=np.zeros([N],dtype=int)
# state2 will store the shifted quantum state
state2=np.zeros([2**N],dtype=np.complex_)
# Next for loop calculates, T_L|ψ⟩ or T_R|ψ⟩, as per choice
for i in range(0,vec.shape[0]):
    if vec[i] != complex(0.0,0.0):
        basis=self.decimal_binary(i,N)
        basis=self.binary_shift(basis,shift=shift,\
                               shift_direction=shift_direction)
        j=self.binary_decimal(basis)
        state2[j]=vec[i]

    return state2
```

Example

In this example, we right shift the following quantum state by 1 position to the left

$$|\psi\rangle = (0.141776 + 0.212664i)|001\rangle + (0.283552 + 0.354441i)|010\rangle +$$
$$(0.567104 + 0.637993i)|100\rangle.$$

```python
A = np.array([0, complex(0.141776 , 0.212664), complex(0.283552 , \
            0.354441),0, complex(0.567104 , 0.637993), 0, 0, 0])
B=quantum_obj.cstate_shift(A,shift=1,shift_direction='right')
f = open('shifted_state.txt','w')
for row in B:
    f.write(str(row))
    f.write('\n')
f.close()
```

The entries in the text file 'shifted_state.txt' is shown below:

```
0j
(0.283552+0.354441j)
(0.567104+0.637993j)
0j
(0.141776+0.212664j)
0j
0j
0j
```

2.12   Complete example

In this section, we provide mathematical description of a quantum mechanical problem, thereafter we show the complete Python code for the benefit of the readers. We first create a two-qubit non-normalized quantum state as shown below.

$$|\psi_1\rangle = \begin{pmatrix} 1 & 2 & 3 & 4 \end{pmatrix}^T. \tag{2.27}$$

In order to normalize the state in Eq. [2.27], we will use the standard prescription,

$$|\widetilde{\psi_1}\rangle = \frac{|\psi_1\rangle}{\sqrt{\langle\psi_1|\psi_1\rangle}}. \tag{2.28}$$

Thereafter, we form the outer product of the state $|\widetilde{\psi_1}\rangle$ in order to create the pure state density matrix ρ_1 as shown below.

$$\rho_1 = |\widetilde{\psi_1}\rangle\langle\widetilde{\psi_1}|. \tag{2.29}$$

As another example, we create a single qubit normalized quantum state shown below.

$$|\psi_2\rangle = \left(\frac{1}{\sqrt{2}} \quad \frac{1}{\sqrt{2}}\right)^T. \tag{2.30}$$

We now take the tensor product of the state $|\psi_2\rangle\langle\psi_2|$ with the density matrix in Eq. [2.29] to obtain another density matrix ρ as,

$$\rho = \rho_1 \otimes \rho_2 = |\widetilde{\psi_1}\rangle\langle\widetilde{\psi_1}| \otimes |\psi_2\rangle\langle\psi_2|. \tag{2.31}$$

The code for the above quantum task is shown below.

```python
import numpy as np
from QuantumInformation import QuantumMechanics as QM

quantum_obj=QM()

# Constructing the state |ψ₁⟩
v1=np.array([1.0,2.0,3.0,4.0])
# Normalizing the state |ψ₁⟩
v1=quantum_obj.normalization_vec(v1)
# Creating the pure state density matrix ρ₁
rho_1=quantum_obj.outer_product_rvec(v1,v1)
# Constructing the state |ψ₂⟩
v2=np.array([1.0/np.sqrt(2),1/np.sqrt(2)])
# Finally constructing the density matrix ρ = ρ₁ ⊗ ρ₂
rho=quantum_obj.tensor_product_matrix(rho_1,\
                    quantum_obj.outer_product_rvec(v2,v2))

print("Trace of the final density matrix =",np.matrix.trace(rho))
```

The above code generate prints the following result

```
Trace of the final density matrix = 0.999999999999997
```

3

Numerical Linear Algebra Operations

Having done with the basic recipes involving vector and matrix elements in the previous chapter, here in this chapter we will give the recipes for some major linear algebra operations [26, 29, 30]. These operations are very important from the point of view of many quantum information tasks and also required in the solution of major problems in quantum mechanics and elsewhere. To be more specific, this chapter broadly deals with matrix decompositions, calculation of different matrix norms, functions of matrices and orthogonalization procedures [25, 31]. However, in this chapter, we have purposely not included a few numerical methods, which have been included in our earlier book. The reason behind such a decision is because the NumPy library already included those methods, and therefore it is redundant to include those methods in our **QuantumInformation** library. The methods which are present in the NumPy library and not included in our library are listed below.

Method	Description
numpy.linalg.svd	Computes singular value decomposition of a matrix.
numpy.kron	Performs Kronecker product of two arrays.
numpy.linalg.qr	Computes the QR decomposition of a matrix.

Linear algebra is a branch of mathematics that deals with vectors, vector spaces, linear maps, and a system of linear equations. However, we will be studying those topics which are pertaining to quantum information theory. All the methods which will be discussed in this chapter are written inside the class **LinearAlgebra** and this whole class is written inside the Python module **chap3_linearalgebra.py**. The methods inside the the class **LinearAlgebra** also uses some functions from the NumPy, Math, Cmath, and SciPy libraries. In the forthcoming sections, we will use the following matrices for the purpose of illustration of how recipes work:

For a real symmetric matrix,

$$A_1 = \begin{pmatrix} 1 & -2 & 3 \\ -2 & 3 & 4 \\ 3 & 4 & 5 \end{pmatrix}. \tag{3.1}$$

For a complex Hermitian matrix,

$$A_2 = \begin{pmatrix} 2 & 1 - 3i \\ 1 + 3i & 4 \end{pmatrix}. \tag{3.2}$$

DOI: 10.1201/9781003285489-3

For a real non-symmetric matrix,

$$A_3 = \begin{pmatrix} 1 & -2 & 3 \\ 2 & 3 & 5 \\ -4 & 4 & 5 \end{pmatrix}. \tag{3.3}$$

For a complex non-symmetric matrix,

$$A_4 = \begin{pmatrix} 2 & 1 - 3i \\ 2 + i & 4 \end{pmatrix}. \tag{3.4}$$

To import the class **LinearAlgebra** from the **QuantumInformation** library, and thereafter we have to create an object of the class. The coding of the preceding statement can be done as follows,

```
# importing the class LinearAlgebra from the QuantumInformation library
from QuantumInformation import LinearAlgebra as LA

# creating the object of the class
linear_obj=LA()
```

3.1 Inverse of a matrix

For a square matrix, A if there exists another matrix B such that $AB = BA = I$, then matrix B is called the inverse of A. The inverse of A is defined as follows A^{-1}. For an inverse of a matrix A to exist, it must satisfy the following conditions,

- The determinant of the matrix A should be non-zero, i.e. $det(A) \neq 0$.

- The matrix A must be square.

The inverse of the matrix A can be obtained from the following mathematical expression,

$$A^{-1} = \frac{adj(A)}{det(A)}, \tag{3.5}$$

where $adj(A)$ is the transpose of its co-factor matrix.

3.1.1   For a real or complex matrix

```
linear_obj.inverse_matrix(A)
```

Parameters

in/out	Argument	Description
[in]	mat	It is a real or complex array of dimension $(0{:}n{-}1, 0{:}n{-}1)$, which is the given matrix
[out]	B	B is an array of dimension $(0{:}n{-}1, 0{:}n{-}1)$, which is the inverse of the matrix mat.

Implementation

```python
def inverse_matrix(self,mat):
    """ Calculates the inverse of a matrix
        Attributes:
            mat : Inverse of the array or matrix to be calculated.
        Return: inverse of matrix mat
    """
    # Here we check the determinant of the matrix det(A) != 0
    assert np.linalg.det(mat) != 0, "Determinant of the matrix is zero"
    return np.linalg.inv(mat)
```

Example

In this example, we will find the inverse of the real matrix given below:

$$A = \begin{pmatrix} 1 & 2 & 3 \\ 3 & -2 & -5 \\ 7 & -2 & -3 \end{pmatrix}. \tag{3.6}$$

```python
import numpy as np
A=np.array([[1, 2, 3], [3, -2, -5], [7, -2, -3]])
A_inv=linear_obj.inverse_matrix(A)
print(A_inv)
```

The above code prints the following output,

```
[[ 0.125    0.      0.125 ]
 [ 0.8125   0.75   -0.4375]
 [-0.25    -0.5     0.25  ]]
```

Similarly, we can find the inverse of a complex matrix given below:

$$A = \begin{pmatrix} 1+i & 1+i \\ 3-3i & 4-i \end{pmatrix}. \tag{3.7}$$

```python
A=np.array([[complex(1,1), complex(1,1)], [complex(3,-3), complex(4,-1)]])
A_inv=linear_obj.inverse_matrix(A)
print(A_inv)
```

The above code prints the following output,

```
[[-0.7-1.1j  -0.2+0.4j]
 [ 1.2+0.6j   0.2-0.4j]]
```

3.2   Function of a matrix

Given any square matrix M of dimension $n \times n$, we can ask the question, what will be the sine of the matrix? [32] or what will be the power of the matrix? It is known that any real continuous function $f(x)$ (or a complex analytic function $f(z)$) has a Taylor series expansion around a point a given by,

$$f(x) = \sum_{n=0}^{\infty} \frac{f^{(n)}(a)}{n!}(x-a)^n, \tag{3.8}$$

if the expansion is done around $a = 0$, it is called a Maclaurin expansion which has the form,

$$f(x) = \sum_{n=0}^{\infty} \frac{f^{(n)}(0)}{n!}x^n. \tag{3.9}$$

Just like any function above, any function of matrix M also can be expanded using the Maclaurin expansion as follows,

$$f(M) = \mathbb{I}f(0) + \frac{f^1(0)}{1!}M + \frac{f^2(0)}{2!}M^2 + \frac{f^3(0)}{3!}M^3 + \cdots = \sum_{n=0}^{\infty} \frac{f^n(0)}{n!}M^n. \tag{3.10}$$

If the matrix is diagonalizable, the above equation reduces to a matrix form given as below,

$$f(M) = Pf(D)P^{-1}, \tag{3.11}$$

where P is a matrix, whose columns are made of the normalized eigenvectors of M and D, is a diagonal matrix which contains the eigenvalues of M as diagonal elements. If M is a normal matrix, Eq. [3.10] is reduced to

$$f(M) = Uf(D)U^{\dagger}, \tag{3.12}$$

where U is a unitary matrix. As we know almost all the problems of physics and specially in quantum mechanics and quantum information theory involves only Hermitian matrices, they are "compulsorily" diagonalizable, because every Hermitian matrix is a normal matrix which can be diagonalized by a unitary transform. Thus Eq. [3.12] can be used to find the function of matrices. We require the exponential of a matrix when we study the time evolution of quantum systems. In finding quantum entropies, we require the logarithm of a matrix and in many other diverse areas of other functions of matrices like power and trigonometric functions are used. Note that for matrices which are not diagonalizable, these techniques can not be used and in such cases the power series with proper truncation is used based on the accuracy of the results we require.

3.2.1   For a real symmetric matrix

```
function_smatrix(mat1,mode,log_base)
```

Parameters

in/out	Argument	Description
[in]	mat1	It is the symmetric matrix of dimension $(0{:}n{-}1,0{:}n{-}1)$ whose function of the matrix is to be calculated
[in]	mode	It defines the type of function of matrix. The allowed types of functions are shown below; if mode='exp', it will compute exp(A). It is the default mode if mode='sin', it will compute sin(A) if mode='cos', it will compute cos(A) if mode='tan', it will compute tan(A) if mode='log', it will compute log(A)
[in]	log_base	It stores the base of the log function. The default value is equal to 2
[out]	B	It is any array of dimension $(0{:}n{-}1,0{:}n{-}1)$, which stores the function of matrix mat1

Implementation

```python
def function_smatrix(self, mat1, mode="exp",log_base=2):
    """
    It calculates the function of a real symmetric matrix.
    Attributes:
        mat1 : The symmetric matrix of which function is to be calculated.
        mode: Primarily calculates the following,
            mode='exp': Exponential of a matrix. It is the default mode.
            mode='sin': sine of a matrix.
            mode='cos': cosine of matrix.
            mode='tan': tan of matrix.
            mode='log': Logarithm of a matrix, by default log base 2.
        log_base: base of the log function
    Return: Function of symmetric matrix mat1
    """
    # Checking whether the matrix is symmetric or not, M^T = M
    assert np.allclose(mat1, np.matrix.transpose(mat1))==True,\
        "The matrix entered is not a symmetric matrix"

    # Checking whether the matrix is a square matrix or not, M_{n×n}
    assert mat1.shape[0] == mat1.shape[1],\
        "Entered matrix is not a square matrix"

    # Checking whether the entered mode is valid or not
    if mode not in ["exp","sin","cos","tan","log"]:
        raise Exception(f"Sorry, the entered mode {mode} is not available")
```

```
eigenvalues,eigenvectors,info=la.dsyev(mat1)

if mode == 'exp':
    diagonal=np.zeros((mat1.shape[0],mat1.shape[1]),dtype=float)
    # Constructing e^D
    for i in range(0,diagonal.shape[0]):
        diagonal[i,i] = math.exp(eigenvalues[i])

if mode == 'sin':
    diagonal=np.zeros((mat1.shape[0],mat1.shape[1]),dtype=float)
    # Constructing sin(D)
    for i in range(0,diagonal.shape[0]):
        diagonal[i,i] = math.sin(eigenvalues[i])

if mode == 'cos':
    diagonal=np.zeros((mat1.shape[0],mat1.shape[1]),dtype=float)
    # Constructing cos(D)
    for i in range(0,diagonal.shape[0]):
        diagonal[i,i] = math.cos(eigenvalues[i])

if mode == 'tan':
    diagonal=np.zeros((mat1.shape[0],mat1.shape[1]),dtype=float)
    # Constructing tan(D)
    for i in range(0,diagonal.shape[0]):
        diagonal[i,i] = math.tan(eigenvalues[i])

if mode == 'log':
    diagonal=np.zeros((mat1.shape[0],mat1.shape[1]),dtype=float)
    # Constructing log(D)
    for i in range(0,diagonal.shape[0]):
        # Checking the eigenvalues of M to be greater than 0.
        assert eigenvalues[i] > 0.0,\
        "eigenvalues of the matrix are negative or zero"
        diagonal[i,i] = math.log(eigenvalues[i],log_base)

# Finally we return, Of(D)O^T, where O is an orthogonal matrix
return np.matmul(np.matmul(eigenvectors,diagonal),\
                 np.matrix.transpose(eigenvectors))
```

Example

In this example, we are finding the exponential of the matrix A_1.

```
A=np.array([[1, -2, 3], [-2, 3, 4], [3, 4, 5]])
B=linear_obj.function_smatrix(A,"exp")
print(B)
```

The above code prints the exponential of the matrix A_1.

```
[[ 168.28088999   404.54130675   646.4572661 ]
 [ 404.54130675  1251.99678938  1854.11449223]
 [ 646.4572661   1854.11449223  2805.68864861]]
```

Now we show how robust is our function **function_smatrix** in detecting errors, we present the following case. Consider the following real symmetric matrix

$$\Phi = \begin{pmatrix} -1 & 0 & 0 \\ 0 & 2 & 11 \\ 0 & 11 & 5 \end{pmatrix}. \tag{3.13}$$

The eigenvalues of the matrix in Eq. [3.13] can be simply calculated to be -1, -7.6018 and 14.6018. The matrix Φ has two negative eigenvalues, therefore, we cannot calculate $\ln(\Phi)$. If we try to calculate $\ln(\Phi)$ using **function_smatrix** then it would prompt error message as shown below.

```
matrix = np.array([[-1, 0, 0],[0, 2, 11],[0, 11, 5]])
B=linear_obj.function_smatrix(matrix,"log")
```

The above code generates the following error message.

```
AssertionError: eigenvalues of the matrix are negative or zero
```

3.2.2   For a complex Hermitian matrix

```
function_hmatrix(mat1, mode,log_base)
```

Parameters

in/out	Argument	Description
[in]	mat1	It is the Hermitian matrix of dimension $(0:n-1,0:n-1)$ whose function of the matrix is to be calculated
[in]	mode	It defines the type of function of matrix. The allowed types of functions are shown below if mode='exp', it will compute exp(A). It is the default mode if mode='sin', it will compute sin(A) if mode='cos', it will compute cos(A) if mode='tan', it will compute tan(A) if mode='log', it will compute log(A)
[in]	log_base	It stores the base of the log function. The default value is equal to 2
[out]	B	It is any array of dimension $(0:n-1,0:n-1)$, which stores the function of matrix mat1

Implementation

```python
def function_hmatrix(self, mat1, mode="exp",log_base=2):
    """

    It calculates the function of Hermitian matrix.
    Attributes:
        mat1 : The Hermitian matrix of which function is to be calculated.
        mode: Primarily calculates the following,
            mode='exp': Exponential of a matrix. It is the default mode.
            mode='sin': sine of a matrix.
            mode='cos': cosine of matrix.
            mode='tan': tan of matrix.
            mode='log': Logarithm of a matrix, by default log base 2.
        log_base: base of the log function
    Return: Function of Hermitian matrix mat1
    """


    # Checking Hermiticity of the matrix, $M = M^\dagger$
    assert np.allclose(mat1, np.transpose(np.conjugate(mat1)))==True \
                        ,"The matrix entered is not a hermitian matrix"

    # Checking whether the matrix is a square matrix, $M_{n \times n}$
    assert mat1.shape[0] == mat1.shape[1],\
    "Entered matrix is not a square matrix"

    # Checking the entered mode is valid or not
    if mode not in ["exp","sin","cos","tan","log"]:
        raise Exception(f"Sorry, the entered mode {mode} is not available")

    eigenvalues,eigenvectors,info=la.zheev(mat1)

    if mode == 'exp':
        diagonal=np.zeros((mat1.shape[0],mat1.shape[1]),dtype=float)
        # Constructing $e^D$
        for i in range(0,diagonal.shape[0]):
            diagonal[i,i] = math.exp(eigenvalues[i])

    if mode == 'sin':
        diagonal=np.zeros((mat1.shape[0],mat1.shape[1]),dtype=float)
        # Constructing $\sin(D)$
        for i in range(0,diagonal.shape[0]):
            diagonal[i,i] = math.sin(eigenvalues[i])

    if mode == 'cos':
        diagonal=np.zeros((mat1.shape[0],mat1.shape[1]),dtype=float)
        # Constructing $\cos(D)$
        for i in range(0,diagonal.shape[0]):
            diagonal[i,i] = math.cos(eigenvalues[i])
```

```
if mode == 'tan':
    diagonal=np.zeros((mat1.shape[0],mat1.shape[1]),dtype=float)
    # Constructing tan(D)
    for i in range(0,diagonal.shape[0]):
        diagonal[i,i] = math.tan(eigenvalues[i])

if mode == 'log':
    diagonal=np.zeros((mat1.shape[0],mat1.shape[1]),dtype=float)
    # Constructing log(D)
    for i in range(0,diagonal.shape[0]):
        assert eigenvalues[i] > 0.0, "eigenvalues of the matrix are
negative"
        diagonal[i,i] = math.log(eigenvalues[i],log_base)

# Finally return, Uf(D)U†
return np.matmul(np.matmul(eigenvectors,diagonal),\
                np.transpose(np.conjugate(eigenvectors)))
```

Example

In this example, we are finding $\tan(A_2)$.

```
A= np.array([[2,complex(1,-3)],[complex(1,3),4]])
B=linear_obj.function_hmatrix(A,"tan")
print(B)
```

The above code prints the following output.

```
[[-0.2015357 +0.j          0.05443782-0.16331346j]
 [ 0.05443782+0.16331346j -0.09266005+0.j          ]]
```

3.2.3   For a real or complex non-Hermitian matrix

```
function_gmatrix(mat1, mode="exp",log_base=2)
```

Parameters

in/out	Argument	Description
[in]	mat1	It is a real or complex array of dimension $(0:n-1,0:n-1)$
[in]	mode	It defines the type of function of matrix. The allowed types of functions are shown below if mode='exp', it will compute exp(A). It is the default mode. if mode='sin', it will compute sin(A) if mode='cos', it will compute cos(A) if mode='tan', it will compute tan(A) if mode='log', it will compute log(A)
[in]	log_base	It stores the base of the log function. The default value is equal to 2
[out]	B	It is any array of dimension $(0:n-1,0:n-1)$, which stores the function of matrix mat1

Implementation

```
def function_gmatrix(self, mat1, mode="exp",log_base=2):
    """
    It calculates the function of general diagonalizable matrix.
    Attributes:
        mat1: The general matrix of which function is to be calculated.
        mode: Primarily calculates the following,
                Primarily calculates the following,
                mode='exp': Exponential of a matrix.
                mode='sin': sine of a matrix.
                mode='cos': cosine of matrix.
                mode='tan': tan of matrix.
                mode='log': Logarithm of a matrix, by default log base 2.
    Return: Function of general matrix mat1
    """

    # Checking whether the matrix is a square matrix.
    assert mat1.shape[0] == mat1.shape[1],\
        "Entered matrix is not a square matrix"

    # Checking if entered mode is valid or not.
    if mode not in ["exp","sin","cos","tan","log"]:
        raise Exception(f"Sorry, the entered mode {mode} is not available")

    eigenvalues,eigenvectors=np.linalg.eig(mat1)

    if mode == 'exp':
        diagonal=np.zeros((mat1.shape[0],mat1.shape[1]),dtype=complex)
        # Constructing $e^D$
```

```
        for i in range(0,diagonal.shape[0]):
            diagonal[i,i] = cmath.exp(eigenvalues[i])

    if mode == 'sin':
        diagonal=np.zeros((mat1.shape[0],mat1.shape[1]),dtype=complex)
        # Constructing sin(D)
        for i in range(0,diagonal.shape[0]):
            diagonal[i,i] = cmath.sin(eigenvalues[i])

    if mode == 'cos':
        diagonal=np.zeros((mat1.shape[0],mat1.shape[1]),dtype=complex)
        # Constructing cos(D)
        for i in range(0,diagonal.shape[0]):
            diagonal[i,i] = cmath.cos(eigenvalues[i])

    if mode == 'tan':
        diagonal=np.zeros((mat1.shape[0],mat1.shape[1]),dtype=complex)
        # Constructing tan(D)
        for i in range(0,diagonal.shape[0]):
            diagonal[i,i] = cmath.tan(eigenvalues[i])

    if mode == 'log':
        diagonal=np.zeros((mat1.shape[0],mat1.shape[1]),dtype=complex)
        # Constructing log(D)
        for i in range(0,diagonal.shape[0]):
            diagonal[i,i] = cmath.log(eigenvalues[i],log_base)

    assert np.linalg.det(eigenvectors) != 0, "Determinant of eigenvectors \
                            matrix is zero"
    # Finally return, Pf(D)P⁻¹
    return np.matmul(np.matmul(eigenvectors,diagonal),\
                    np.linalg.inv(eigenvectors))
```

Example

In this example, we are finding the natural log of the matrix A_3 i.e. $\ln(A_3)$.

```
A=np.array([[1, -2, 3], [2, 3, 5],[-4, 4, 5]])
B=linear_obj.function_gmatrix(A,mode='log',log_base=math.exp(1))
print(B)
```

The above code prints the natural log of the matrix A_3 as shown below,

```
[[ 1.64437167+2.72787460e-17j -0.96712394-2.35900820e-16j
   1.01814825-2.38386296e-17j]
 [ 1.21428832+6.75553539e-17j  1.25209154-1.82545025e-16j
   0.68765921-3.30220382e-17j]
 [-1.19275475-3.99432226e-17j  0.45126162+1.58309757e-16j
   1.84846891-4.37110136e-17j]]
```

Another example we are finding $\cos(A_4)$.

```
A=np.array([[2, complex(1,-3)],[complex(2,1),4]])
B=linear_obj.function_gmatrix(A,mode='cos')
print(B)
```

The above code prints the cosine of the matrix A_4.

```
[[ 1.3008916 -0.47788983j -0.19258167-0.00342889j]
 [ 0.02165839-0.13446428j  1.2644326 -0.59412461j]]
```

3.3   Power of a matrix

Consider a square matrix A and positive number k, the kth power of the matrix A is defined as follows A^k. If the inverse of the matrix A exist, then A^{-k} can be calculated. The power of a square matrix A can be found using the spectral theorem as follows. Let us consider that the matrix A is diagonalizable with matrix D containing the eigenvalues of A along main diagonal, that is $D = P^{-1}AP$, this implies $A = PDP^{-1}$. Multiplying these equations k times and using the fact that $PP^{-1} = P^{-1}P = \mathbb{I}$, we can see that,

$$A^k = PD^kP^{-1}. \tag{3.14}$$

Note that in the above equations if A is a Hermitian matrix, $A^k = UD^kU^\dagger$, where U is a unitary matrix. Using Eq. [3.14], we can also calculate power of the matrix $A^{k'}$, where k' is a irrational number.

3.3.1   For a real symmetric matrix

```
power_smatrix(mat1,k,precision=10**(-10))
```

Parameters

in/out	Argument	Description
[in/out]	mat1	mat1 is a real array of dimension (0:n−1,0:n−1)
[in]	k	k is the power to which matrix A is raised
[in]	precision	If the absolute value of any eigenvalues of mat1 is below precision, then it will consider those particular eigenvalues to zero. The default value is 10^{-10}
[out]	B	B is a complex or real array of dimension (0:n−1,0:n−1), which will store the kth power of matrix mat1

Implementation

```
def power_smatrix(self,mat1,k,precision=10**(-10)):
    """
    It calculates the power of a real symmetric matrix.
    Attributes:
        mat1 : The matrix or array of which power is  to be calculated.
        k : value of the power
        precision: if the absolute eigenvalues below the precision
                    value will be considered as zero
    Return: k'th Power of symmetric matrix mat1
    """
    eigenvalues,eigenvectors,info=la.dsyev(mat1)
    flag=0
    for i in eigenvalues:
        # Checking any eigenvalues is negative
        if i < 0.0:
            flag=1
    # If all eigenvalues are +ve, then D is real type
    if flag==0:
        diag=np.zeros([eigenvectors.shape[0],eigenvectors.shape[1]],\
                    dtype='float64')
    # If any eigenvalues is -ve, then D is complex type
    else:
        diag=np.zeros([eigenvectors.shape[0],eigenvectors.shape[1]],\
                    dtype='complex_')
    # Constructing D^k
    for i in range(0,eigenvectors.shape[0]):
        if abs(eigenvalues[i]) <= precision:
            diag[i,i]=0.0
            eigenvalues[i]=0.0
        if eigenvalues[i] < 0.0:
            diag[i,i]=pow(abs(eigenvalues.item(i)),k)*pow(complex(0,1),2*k)
        else:
            diag[i,i]=pow(eigenvalues.item(i),k)
    # Constructing OD^kO^T, where O is orthogonal matrix
    diag=np.matmul(np.matmul(eigenvectors,diag),np.transpose(eigenvectors))
    return diag
```

Example

In this example, we are finding the matrix $A_1^{2/3}$.

```
A=np.array([[1, -2, 3], [-2, 3, 4], [3, 4, 5]])
B=linear_obj.power_smatrix(A,2/3)
print(B)
```

The above code prints $A_1^{2/3}$.

```
[[ 0.99390456+0.77939749j  -1.08157018+0.66289919j   1.42855357-0.61766176j]
 [-1.08157018+0.66289919j   1.84533521+0.56381414j   1.75131174-0.52533846j]
 [ 1.42855357-0.61766176j   1.75131174-0.52533846j   2.62409023+0.48948842j]]
```

3.3.2   For a complex Hermitian matrix

---

power_hmatrix(mat1,k,precision=10**(-10))

Parameters

in/out	Argument	Description
[in]	mat1	mat1 is a complex array of dimension $(0{:}n{-}1,0{:}n{-}1)$
[in]	k	k is the power to which matrix mat1 is raised
[in]	precision	If the absolute value of any eigenvalues of mat1 is below precision, then it will consider those particular eigenvalues to zero. The default value is 10^{-10}
[out]	B	B is a complex or real array of dimension $(0{:}n{-}1,0{:}n{-}1)$ which will store the kth power of matrix mat1

Implementation

```
def power_hmatrix(self,mat1,k,precision=10**(-10)):
    """
    It calculates the power of a Hermitian matrix.
    Attributes:
        mat1 : The matrix or array of which power is  to be calculated.
        k : value of the power
        precision: if the absolute eigenvalues below the precision
                    value will be considered as zero
    Return: k'th Power of Hermitian matrix mat1
    """
    eigenvalues,eigenvectors,info=la.zheev(mat1)
    flag=0
    for i in eigenvalues:
        # Checking any eigenvalues is negative
        if i < 0.0:
            flag=1
    # If all eigenvalues are +ve then D is real type
    if flag==0:
        diag=np.zeros([eigenvectors.shape[0],eigenvectors.shape[1]],\
                    dtype='float64')
    # If any eigenvalue is -ve then D is complex type
    else:
        diag=np.zeros([eigenvectors.shape[0],eigenvectors.shape[1]],\
                    dtype='complex_')
    # Constructing D^k
    for i in range(0,eigenvectors.shape[0]):
        if abs(eigenvalues[i]) <= precision:
            diag[i,i]=0.0
            eigenvalues[i]=0.0
```

```
        if eigenvalues[i] < 0.0:
            diag[i,i]=pow(abs(eigenvalues.item(i)),k)*pow(complex(0,1),2*k)
        else:
            diag[i,i]=pow(eigenvalues.item(i),k)
    # Finally calculate, UD^kU†, where U is unitary matrix
    diag=np.matmul(np.matmul(eigenvectors,diag),np.conjugate(\
                np.transpose(eigenvectors)))
    return diag
```

Example

In this example, we are finding the matrix $A_2^{2/3}$.

```
A=np.array([[2,complex(1,-3)],[complex(1,3),4]])
B=linear_obj.power_hmatrix(A,2/3)
print(B)
```

The above code prints $A_2^{2/3}$.

```
[[1.04224674+0.26180536j  0.36821119-1.71113837j]
 [0.73211407+1.58983741j  2.142572+0.1405044j ]]
```

3.3.3   For a real or complex non-Hermitian matrix

```
power_gmatrix(mat1, k, precision=10**(-10))
```

Parameters

in/out	Argument	Description
[in]	mat1	mat1 is a complex or real array of dimension $(0{:}n{-}1,0{:}n{-}1)$
[in]	k	k is the power to which matrix mat1 is raised
[in]	precision	If the absolute value of any eigenvalues of mat1 is below precision, then it will consider those particular eigenvalues to zero. The default value is 10^{-10}
[out]	B	B is a complex array of dimension $(0{:}n{-}1,0{:}n{-}1)$, which will store the kth power of matrix mat1

Implementation

```
def power_gmatrix(self,mat1, k, precision=10**(-10)):
    """

    Calculates the power of a general non-Herimitian matrix
    Attributes:
        mat1 : The matrix or array of which power is  to be calculated.
        k : value of the power
```

```
        precision: if the absolute eigenvalues below the precision
                    value will be considered as zero
    Return: k'th Power of non-Hermitian matrix mat1
    """
    eigenvalues,eigenvectors=np.linalg.eig(mat1)
    diag=np.zeros([eigenvectors.shape[0],eigenvectors.shape[1]],dtype=np.
    complex_)
    for i in range(0,eigenvectors.shape[0]):
        if abs(eigenvalues[i]) <= precision:
            diag[i,i]=complex(0.0,0.0)
        else:
            diag[i,i]=pow(eigenvalues.item(i),k)
    diag=np.matmul(np.matmul(eigenvectors,diag),\
                        np.linalg.inv(eigenvectors))
    return diag
```

Example

In this example, we are finding the matrix A_3^2.

```
A=np.array([[1, -2, 3], [2, 3, 5],[-4, 4, 5]])
B=linear_obj.power_gmatrix(A,2)
print(B)
```

The above code prints, A_3^2.

```
[[-15.+7.01916036e-16j    4.+2.37837774e-15j    8.-1.50431364e-15j]
 [-12.-2.58040737e-15j   25.+4.92070678e-15j   46.-5.33922109e-15j]
 [-16.+1.56774552e-18j   40.+4.05915425e-15j   33.-1.61071324e-15j]]
```

Another example where we are finding the matrix A_4^2.

```
A=np.array([[2, complex(1,-3)],[complex(2,1),4]])
B=linear_obj.power_gmatrix(A,2)
print(B)
```

The above code prints, A_4^2.

```
[[ 9. -5.j  6.-18.j]
 [12. +6.j 21.  -5.j]]
```

3.4   Trace norm of a matrix

Let A be a $n \times n$ matrix, note that, even if matrix A is not diagonalizable, $A^\dagger A$ is a positive normal matrix and hence diagonalizable by a unitary transformation. We then define the trace norm of matrix [25, 33] A as

$$\| A \| = Tr\left(\sqrt{A^\dagger A}\right). \tag{3.15}$$

In the above equation, the square root is uniquely determined by spectral theorem as given by Eq. [3.12]. Trace norm has applications in the calculation of logarithmic negativity of a state, computation of fidelity between two density matrices and so on.

3.4.1 For a real matrix

```
trace_norm_rmatrix(mat1, precision=10**(-13))
```

Parameters

in/out	Argument	Description
[in]	mat1	It is a real array of dimension $(0{:}n{-}1, 0{:}n{-}1)$
[in]	precision	If the absolute value of any eigenvalues of mat1 is below precision, then it will consider those particular eigenvalues to zero. The default value is 10^{-13}
[out]	trace_norm	trace_norm is the trace norm of the matrix mat1

Implementation

```
def trace_norm_rmatrix(self, mat1, precision=10**(-13)):
    """
    Calculates the trace norm of a real matrix
    Attributes:
        mat1 : The matrix or array of which trace norm is to be calculated.
        precision: the absolute value of the eigenvalues below precision
                   value will be considered as zero
    Return:
        trace_norm: trace norm of matrix mat1
    """
    # Calculating eigenvalues of A^T A
    eigenvalues,eigenvectors,info=la.dsyev(np.matmul(np.transpose(mat1),\
                                            mat1))
    trace_norm=0.0
    # Here we calculate Tr(sqrt(A^T A))
    for i in range(len(eigenvalues)):
        if abs(eigenvalues[i]) < precision:
            eigenvalues[i]=0.0
        trace_norm=trace_norm+np.sqrt(eigenvalues[i])

    return trace_norm
```

Example

In this example, we are finding the trace norm of the matrix A_3.

```
A=np.array([[1, -2, 3], [2, 3, 5],[-4, 4, 5]])
trace_norm=linear_obj.trace_norm_rmatrix(A)
print(trace_norm)
```

The above code prints the trace norm of the matrix A_3.

```
16.358044723373784
```

3.4.2   For a complex matrix

```
trace_norm_cmatrix(mat1, precision=10**(-13))
```

Parameters

in/out	Argument	Description
[in]	mat1	It is a complex array of dimension $(0{:}n{-}1,0{:}n{-}1)$
[in]	precision	If the absolute value of any eigenvalues of mat1 is below precision, then it will consider those particular eigenvalues to zero. The default value is 10^{-13}
[out]	trace_norm	trace_norm is the trace norm of the matrix mat1

Implementation

```
def trace_norm_cmatrix(self,mat1, precision=10**(-13)):
    """
    Calculates the trace norm of a complex matrix
    Attributes:
        mat1 : The matrix or array of which trace norm is to be calculated.
        precision: the absolute value of the eigenvalues below precision
                    value will be considered as zero.
    Return:
        trace_norm: trace norm of matrix mat1
    """
    # Calculating eigenvalues of A†A
    eigenvalues,eigenvectors,info=\
    la.zheev(np.matmul(np.conjugate(np.transpose(mat1)),mat1))
    trace_norm=0.0
    # Calculating Tr(√A†A)
    for i in range(len(eigenvalues)):
        if abs(eigenvalues[i]) < precision:
            eigenvalues[i]=0.0
        trace_norm=trace_norm+np.sqrt(eigenvalues[i])

    return trace_norm
```

Example

In this example, we are finding the trace norm of the matrix A_4.

```
A=np.array([[2, complex(1,-3)],[complex(2,1),4]])
trace_norm=linear_obj.trace_norm_cmatrix(A)
print(trace_norm)
```

The above code prints the trace norm of the matrix A_4.

```
6.8309518948453007
```

3.5   Hilbert-Schmidt norm of a matrix

Let A be a $n \times n$ matrix, the Hilbert-Schmidt norm [25, 33] of matrix A can be defined as,

$$\| A \|_{HS} = \sqrt{Tr(A^\dagger A)}. \tag{3.16}$$

3.5.1   For a real matrix

```
hilbert_schmidt_norm_rmatrix(mat1, precision=10**(-13))
```

Parameters

in/out	Argument	Description
[in]	mat1	mat1 is a real array of dimension (0:n−1,0:n−1)
[in]	precision	If the absolute value of any eigenvalues of mat1 is below precision, then it will consider those particular eigenvalues to zero. The default value is 10^{-13}
[out]	htrace_norm	It is the Hilbert-Schmidt norm of the matrix mat1

Implementation

```
def hilbert_schmidt_norm_rmatrix(self,mat1, precision=10**(-13)):
    """
    Calculates the Hilbert-Schmidt norm of matrix of a real matrix
    Attributes:
        mat1 : The matrix or array of which Hilbert-Schmidt norm
                is to be calculated.
        precision: tolerance value, the magnitude of eigenvalues below
                precision is considered zero
    Return:
        htrace_norm: Hilbert-Schmidt norm of matrix mat1
    """
```

```
# Calculating eigenvalues of A^T A
eigenvalues,eigenvectors,info=la.dsyev(np.matmul(np.transpose(mat1),\
                                       mat1))
htrace_norm=0.0
# Calculating Tr(A^T A)
for i in range(len(eigenvalues)):
    if abs(eigenvalues[i]) < precision:
        eigenvalues[i]=0.0
    htrace_norm=htrace_norm+eigenvalues[i]
# Calculating sqrt(Tr(A^T A))
htrace_norm=np.sqrt(htrace_norm)
return htrace_norm
```

Example

In this example, we are finding the Hilbert-Schmidt norm of the matrix A_3.

```
A=np.array([[1, -2, 3], [2, 3, 5],[-4, 4, 5]])
htrace_norm=linear_obj.hilbert_schmidt_norm_rmatrix(A)
print(htrace_norm)
```

The above code prints Hilbert-Schmidt norm of the matrix A_3.

```
10.44030650891055
```

3.5.2   For a complex matrix

```
hilbert_schmidt_norm_cmatrix(mat1, precision=10**(-13))
```

Parameters

in/out	Argument	Description
[in]	mat1	mat1 is a complex array of dimension $(0{:}n{-}1,0{:}n{-}1)$
[in]	precision	If the absolute value of any eigenvalues of mat1 is below precision, then it will consider those particular eigenvalues to zero. The default value is 10^{-13}
[out]	htrace_norm	It is the Hilbert-Schmidt norm of the matrix mat1

Implementation

```
def hilbert_schmidt_norm_cmatrix(self,mat1, precision=10**(-13)):
    """
    Calculates the trace norm of a complex matrix
    Attributes:
        mat1 : The matrix or array of which Hilbert-Schmidt norm
               is to be calculated.
```

```
          precision: tolerance value, the magnitude of eigenvalues below
                     precision is considered zero.
    Return:
          htrace_norm: Hilbert-Schmidt norm of matrix mat1.
    """
    # Calculating eigenvalues of A†A
    eigenvalues,eigenvectors,info=\
    la.zheev(np.matmul(mat1,np.conjugate(np.transpose(mat1))))
    htrace_norm=0.0
    # Calculating Tr(A†A)
    for i in range(len(eigenvalues)):
        if abs(eigenvalues[i]) < precision:
            eigenvalues[i]=0.0
        htrace_norm=htrace_norm+eigenvalues[i]
    # Calculating √(Tr(A†A))
    htrace_norm=np.sqrt(htrace_norm)

    return htrace_norm
```

Example

In this example, we are finding the Hilbert-Schmidt norm of the matrix A_4.

```
A=np.array([[2, complex(1,-3)],[complex(2,1),4]])
htrace_norm=linear_obj.hilbert_schmidt_norm_cmatrix(A)
print(htrace_norm)
```

The above code prints the Hilbert-Schmidt norm of the matrix A_4.

```
5.9160797830996161
```

3.6 Absolute value of a matrix

Let A be a $n \times n$ matrix, the absolute value of matrix [34] A can be defined as

$$absA = \sqrt{A^\dagger A}, \qquad (3.17)$$

here the square root is the positive square root.

3.6.1 For a real matrix

```
absolute_value_rmatrix(mat1)
```

Parameters

in/out	Argument	Description
[in]	mat1	It is a real array of dimension $(0{:}n-1,0{:}n-1)$
[out]	res_mat	It is a real array of dimension $(0{:}n-1,0{:}n-1)$, which is the absolute value of the matrix mat1

Implementation

```
def absolute_value_rmatrix(self,mat1):
    """

    Calculates the absolute value of a real matrix
    Attributes:
        mat1 : The matrix of which absolute form has to calculated.
    Return:
        res_mat: Absoulte value of matrix mat1
    """
    # Calculating √A^T A
    res_mat=self.power_smatrix(np.matmul(np.transpose(mat1),\
                                    mat1),0.50)

    return res_mat
```

Example

In this example, we are finding the absolute value of the matrix A_3.

```
A=np.array([[1, -2, 3], [2, 3, 5],[-4, 4, 5]])
B=linear_obj.absolute_value_rmatrix(A)
print(B)
```

The above code prints the entries of the absolute value of the matrix A_3.

```
[[ 4.40016286 -1.23216656 -0.34688963]
 [-1.23216656  4.6660871   2.38943439]
 [-0.34688963  2.38943439  7.29179476]]
```

3.6.2   For a complex matrix

```
absolute_value_cmatrix(mat1)
```

Parameters

in/out	Argument	Description
[in]	mat1	It is a complex array of dimension $(0{:}n{-}1,0{:}n{-}1)$
[out]	res_mat	It is a complex array of dimension $(0{:}n{-}1,0{:}n{-}1)$, which is the absolute value of matrix mat1

Implementation

```
def absolute_value_cmatrix(self,mat1):
    """
    Calculates the absolute value of a complex matrix
    Attributes:
        mat1 : The matrix of which absolute form has to calculated.
    Return:
        res_mat: Absoulte value of matrix mat1
    """
    # Calculating √A†A
    res_mat=self.power_hmatrix(np.matmul(np.conjugate(np.transpose(mat1)),\
                               mat1),0.50)

    return res_mat
```

Example

In this example, we are finding the absolute value of the matrix A_4.

```
A=np.array([[2, complex(1,-3)],[complex(2,1),4]])
B=linear_obj.absolute_value_cmatrix(A)
print(B)
```

The above code prints the entries of the absolute value of the matrix A_4.

```
[[2.17113985+0.j         1.46392482-1.46392482j]
 [1.46392482+1.46392482j 4.65981204+0.j          ]]
```

3.7 The Hilbert-Schmidt inner product between two matrices

Let A and B be two $n \times n$ matrices or operators, the Hilbert-Schmidt inner product [35] between them can be defined as,

$$\langle A, B \rangle_{HS} = Tr(A^\dagger B) \tag{3.18}$$

It is also defined as the Frobenius inner product.

3.7.1   For real or complex matrices

`hilbert_schmidt_inner_product(A,B)`

Parameters

in/out	Argument	Description
[in]	A	A is a real or complex array of dimension $(0{:}n-1,0{:}n-1)$
[in]	B	B is a real or complex array of dimension $(0{:}n-1,0{:}n-1)$
[out]	hs_innp	It is the Hilbert-Schmidt inner product between matrices A and B

Implementation

```
def hilbert_schmidt_inner_product(self,A,B):
    """
    Calculates the Hilbert-Schmidt inner product between matrices.
    Attributes:
        A: It is a complex or real input matrix.
        B: It is a complex or real input matrix.
    Return: Hilbert-Schmidt inner product between A and B.
    """
    # Calculating A†B
    return np.trace(np.matmul(np.conjugate(np.transpose(A)),B))
```

Example

In this example, we are finding the Hilbert-Schmidt inner product between matrices A_1 and A_3.

```
A=np.array([[1, -2, 3], [-2, 3, 4], [3, 4, 5]])
B=np.array([[1, -2, 3], [2, 3, 5],[-4, 4, 5]])
hs_innp=linear_obj.hilbert_schmidt_inner_product(A,B)
print(hs_innp)
```

The above code prints the Hilbert-Schmidt inner product between matrices A_1 and A_3.

68

Another example where we are finding the Hilbert-Schmidt inner product between matrices A_2 and A_4.

```
A=np.array([[2,complex(1,-3)],[complex(1,3),4]])
B=np.array([[2, complex(1,-3)],[complex(2,1),4]])
hs_innp=linear_obj.hilbert_schmidt_inner_product(A,B)
print(hs_innp)
```

The above prints the Hilbert-Schmidt inner product between matrices A_2 and A_4.

(35-5j)

3.8 Gram-Schmidt orthogonalization

Let there be a linearly independent non-orthogonal basis set $|w\rangle : \{|w_1\rangle, |w_2\rangle, \cdots, |w_m\rangle\}$ (i.e. $\langle w_i|w_j\rangle \neq \delta_{ij}$) in some vector space V where an inner product is defined. There is a useful method known as "Gram-Schmidt" procedure [24, 36, 37], which can be used to generate a set of orthonormal vectors $|v\rangle : \{|v_1\rangle, |v_2\rangle, \cdots, |v_m\rangle\}$ from the non-orthonormal set $|w\rangle$. Define initially,

$$|v_1\rangle = \frac{|w_1\rangle}{\||w_1\rangle\|}. \tag{3.19}$$

For $1 \leq k \leq m-1$ define $|v_{k+1}\rangle$ inductively by,

$$|v_{k+1}\rangle \equiv \frac{|w_{k+1}\rangle - \sum_{i=1}^{k}\langle v_i|w_{k+1}\rangle|v_i\rangle}{\||w_{k+1}\rangle - \sum_{i=1}^{k}\langle v_i|w_{k+1}\rangle|v_i\rangle\|}. \tag{3.20}$$

The Gram-Schmidt orthogonalization procedure proves to be very helpful in context to quantum mechanical observable (Hermitian operator), which contains degenerate eigenvalues. The eigenvectors of the degenerate eigenvalues are not necessarily to be orthogonal to each other. However, non-degenerate eigenvectors are orthogonal to each other. To transform the complete eigenvectors into orthonormal eigenvectors, we perform the Gram-Schmidt orthogonalization. This transformation is not a unitary transformation of the eigenbasis.

3.8.1 For real vectors

gram_schmidt_ortho_rmatrix(vectors)

Parameters

in/out	Argument	Description
[in]	vectors	It is a real array of dimension (0:n−1,0:k−1), whose columns are the linearly independent non-orthonormal set of vectors
[out]	orthonormal_vec	It is a real array of dimension (0:n−1,0:k−1), which stores the k vectors as columns of dimension n which are the orthonormalized set of vectors

Implementation

```
def gram_schmidt_ortho_rmatrix(self,vectors):
    """
    Orthornormal set of real vectors are calculated
    Attributes:
        vectors: A matrix whose columns are non-orthogonal set real vectors
    Return:
        orthonormal_vec: A matrix whose columns are orthonormal to each other
    """
```

```
    orthonormal_vec=np.zeros((vectors.shape[0],vectors.shape[1]),
                                   dtype='float64')
    for col in range(0,vectors.shape[1]):
        if col != 0:
            for col2 in range(0,col):
                tr=0.0
                # Here we calculate ⟨vᵢ|w_{k+1}⟩
                for row2 in range(0,vectors.shape[0]):
                    tr=tr+(orthonormal_vec[row2,col2]*vectors[row2,col])
                # Here we calculate ∑_{i=1}^{k}⟨vᵢ|w_{k+1}⟩|vᵢ⟩
                orthonormal_vec[:,col]=orthonormal_vec[:,col]+\
                                       (tr*orthonormal_vec[:,col2])
            # Here we calculate |w_{k+1}⟩ - ∑_{i=1}^{k}⟨vᵢ|w_{k+1}⟩|vᵢ⟩
            orthonormal_vec[:,col]=vectors[:,col]-orthonormal_vec[:,col]
        if col == 0:
            # Here calculate |v₁⟩ = |w₁⟩
            orthonormal_vec[:,col]=vectors[:,col]
        tr=0.0
        # Finally we calculate  (|w_{k+1}⟩ - ∑_{i=1}^{k}⟨vᵢ|w_{k+1}⟩|vᵢ⟩) / (|||w_{k+1}⟩ - ∑_{i=1}^{k}⟨vᵢ|w_{k+1}⟩|vᵢ⟩||)
        for row in range(0,vectors.shape[0]):
            tr=tr+(orthonormal_vec[row,col]*orthonormal_vec[row,col])
        orthonormal_vec[:,col]=orthonormal_vec[:,col]/np.sqrt(tr)

    return orthonormal_vec
```

Example

In this example, we are doing the Gram-Schmidt orthogonalization of the matrix A whose columns are the linearly independent non-orthogonal vectors.

$$
A = \begin{pmatrix} 1 & 2 & 3 & 4 \\ 2 & 33 & 4 & 5 \\ 3 & 4 & 53 & 6 \\ 4 & 5 & 6 & 73 \end{pmatrix}. \tag{3.21}
$$

```
A=np.array([[1,2,3,4],[2, 33, 4, 5],[3,4,53,6],[4,5,6,73]])
B=linear_obj.gram_schmidt_ortho_rmatrix(A)
print(B)
```

The above code prints the orthogonal matrix whose columns are the orthonormal vectors.

```
[[ 0.18257419 -0.04712082 -0.10008963 -0.97694849]
 [ 0.36514837  0.93063624 -0.00416827  0.02377968]
 [ 0.54772256 -0.2120437   0.80879782  0.0297246 ]
 [ 0.73029674 -0.29450514 -0.57949183  0.21005383]]
```

3.8.2   For complex vectors

`gram_schmidt_ortho_cmatrix(vectors)`

Parameters

in/out	Argument	Description
[in]	vectors	It is a complex array of dimension $(0{:}n{-}1,0{:}k{-}1)$, whose columns are the linearly independent non-orthonormal set of vectors
[out]	orthonormal_vec	It is a complex array of dimension $(0{:}n{-}1,0{:}k{-}1)$, which stores the k vectors as columns of dimension n which are the orthonormalized set of vectors.

Implementation

```
def gram_schmidt_ortho_cmatrix(self,vectors):
    """
    Orthornormal set of complex vectors are calculated
    Attributes:
      vectors: A matrix whose columns are non-orthogonal set
               complex vectors
    Return:
      orthonormal_vec: A matrix whose columns are orthonormal to each other
    """
    orthonormal_vec=np.zeros((vectors.shape[0],vectors.shape[1]),\
                             dtype=np.complex_)
    for col in range(0,vectors.shape[1]):
        if col != 0:
            # Here initialize, |v_{k+1}> = |w_{k+1}>
            orthonormal_vec[:,col]=vectors[:,col].copy()
            # Next for loop we calculate |v_{k+1}> = |w_{k+1}> - sum_{i=1}^{k}<v_i|w_{k+1}>|v_i>
            for col2 in range(0,col):
                tr=complex(0.0,0.0)
                # Next we calculate <v_i|w_{k+1}>
                for row2 in range(0,vectors.shape[0]):
                    tr=tr+(np.conjugate(orthonormal_vec[row2,col2])\
                        *vectors[row2,col])
                orthonormal_vec[:,col]=orthonormal_vec[:,col]-\
                                        (tr*\
                                        orthonormal_vec[:,col2].copy())
        if col == 0:
            # Here initialize, |v_1> = |w_1>
            orthonormal_vec[:,col]=vectors[:,col].copy()
        tr=complex(0.0,0.0)
```

Finally we calculate $\dfrac{|w_{k+1}\rangle - \sum_{i=1}^{k}\langle v_i|w_{k+1}\rangle|v_i\rangle}{|||w_{k+1}\rangle - \sum_{i=1}^{k}\langle v_i|w_{k+1}\rangle|v_i\rangle||}$

```
    for row in range(0,vectors.shape[0]):
        tr=tr+(np.conjugate(orthonormal_vec[row,col])*\
            orthonormal_vec[row,col])
    orthonormal_vec[:,col]=orthonormal_vec[:,col]/np.sqrt(tr.real)

return orthonormal_vec
```

Example

In this example, we are doing the Gram-Schmidt orthogonalization of the matrix A whose columns are the linearly independent non-orthogonal vectors.

$$A = \begin{pmatrix} 1+i & 2+i \\ 2-i & 4 \end{pmatrix}. \tag{3.22}$$

```
A=np.array([[complex(1,1),complex(2,1)],[complex(2,-1),4]])
B=linear_obj.gram_schmidt_ortho_cmatrix(A)
print(B)
```

The above code prints a unitary matrix, whose columns constructed from the matrix in Eq.[3.22] using Gram-Schmidt orthogonalization process.

```
[[0.37796447+0.37796447j 0.5500191 -0.64168895j]
 [0.75592895-0.37796447j 0.27500955+0.45834925j]]
```

3.9   Complete example

In this section, we first provide mathematical description of a linear algebra problem, thereafter we show the complete Python code which accomplishes the task. To this end, we first construct a symmetric matrix A, as shown below.

$$A = \begin{pmatrix} 1 & 3 & 5 & 7 \\ 3 & 11 & 13 & 19 \\ 5 & 13 & 23 & 29 \\ 7 & 19 & 29 & 31 \end{pmatrix}. \tag{3.23}$$

We calculate the cosine of the matrix in Eq. [3.23].

$$B = \cos(A) = \begin{pmatrix} -0.0281185 & -0.0105118 & 0.0230653 & -0.181293 \\ -0.0105118 & 0.249689 & -0.279703 & -0.383533 \\ 0.0230653 & -0.279703 & -0.0231371 & -0.581786 \\ -0.181293 & -0.383533 & -0.581786 & -0.15813 \end{pmatrix}. \tag{3.24}$$

Finally, we calculate the matrix $B^{3.5}$ as shown below.

$$B^{3.5} = \begin{pmatrix} 0.00714175 - i0.0123718 & 0.00934645 - i0.0345915 & 0.0174189 - i0.0534346 & -0.0211138 - i0.0635809 \\ 0.00934645 - i0.0345915 & 0.0543491 - i0.0968972 & 0.00358065 - i0.149686 & -0.0343968 - i0.178085 \\ 0.0174189 - i0.0534346 & 0.00358065 - i0.149686 & 0.0512519 - i0.231234 & -0.0484105 - i0.275104 \\ -0.0211138 - i0.0635809 & -0.0343968 - i0.178085 & -0.0484105 - i0.275104 & 0.0635073 - i0.3273 \end{pmatrix} \quad (3.25)$$

The code below does these tasks,

```python
import numpy as np
from QuantumInformation import LinearAlgebra as LA
linear_obj=LA()

# Construct the matrix A
A=np.array([[1, 3, 5, 7],[3, 11, 13, 19],\
            [5, 13, 23, 29],[7, 19, 29, 31]])

# Calculating the matrix B
B=linear_obj.function_smatrix(A,"sin")
print("The matrix B obtained")
print(B)

# Calculating the matrix B^3.5
C=linear_obj.power_smatrix(B,3.5)
print("The matrix C obtained")
print(C)
```

The preceding code generates the following code.

```
The matrix B obtained
[[-0.0281185  -0.01051177  0.0230653  -0.18129262]
 [-0.01051177  0.24968853 -0.27970315 -0.3835334 ]
 [ 0.0230653  -0.27970315 -0.0231371  -0.58178568]
 [-0.18129262 -0.3835334  -0.58178568 -0.15812989]]
The matrix C obtained
[[ 0.00714175-0.01237181j   0.00934645-0.03459152j   0.01741888-0.0534346j
  -0.02111381-0.06358091j]
 [ 0.00934645-0.03459152j   0.05434908-0.09689725j   0.00358065-0.14968613j
  -0.03439682-0.17808505j]
 [ 0.01741888-0.0534346j    0.00358065-0.14968613j   0.05125187-0.23123419j
  -0.04841055-0.27510388j]
 [-0.02111381-0.06358091j  -0.03439682-0.17808505j  -0.04841055-0.27510388j
   0.06350731-0.32729968j]]
```

4

Quantum Information and Quantum Computation Tools

In this chapter, we have given the recipes to construct the most important quantum gates [24, 38–41], which are used in quantum computing; however, a suitable combination of these gates can be used to accomplish a specific quantum computing task in hand. Given these gates, they have to act on something, those are the pure states. We have presented how to construct many interesting states which are used in quantum computing. These are the prototypical states in their respective Hilbert spaces. Towards the end, we have given certain quantities like fidelity, trace distance, entropy, as these are important in studying the closeness between two pure states and related to the entanglement content in them. Last but not least, there are recipes dealing with single qubit quantum measurements in the spin basis and expectation values of observables which help us to have a direct mapping between the Hilbert space which resides in the mind of the mathematician to the laboratory which resides in front of us in reality. The following pure states and density matrices will be mostly used to demonstrate the recipes in this chapter.

Real states used for demonstration are,

$$|\psi_1\rangle = \frac{1}{\sqrt{2}}(|00\rangle + |11\rangle), \tag{4.1}$$

$$|\psi_2\rangle = \frac{1}{\sqrt{c_2}}\sum_{x=1}^{2^N} x|x\rangle = \frac{1}{\sqrt{c_2}}\begin{pmatrix} 1 \\ 2 \\ 3 \\ \vdots \\ 2^N \end{pmatrix}, \tag{4.2}$$

where c_2 is the normalization constant given by,

$$c_2 = \sum_{x=1}^{2^N} x^2. \tag{4.3}$$

Complex states used for demonstration are,

$$|\psi_3\rangle = \frac{1}{\sqrt{c_3}}\sum_{x=1}^{2^N}(x - 1 + ix)|x\rangle = \frac{1}{\sqrt{c_3}}\begin{pmatrix} i \\ 1 + 2i \\ 2 + 3i \\ \vdots \\ 2^N - 1 + 2^N i \end{pmatrix}, \tag{4.4}$$

DOI: 10.1201/9781003285489-4

where c_3 is the normalization constant given by,

$$c_3 = \sum_{x=1}^{2^N}((x-1)^2 + x^2),\tag{4.5}$$

$$|\psi_4\rangle = \frac{1}{\sqrt{c_4}}\sum_{x=1}^{2^N}(x-1+i(x+1))|j\rangle = \frac{1}{\sqrt{c_4}}\begin{pmatrix} 2i \\ 1+3i \\ 2+4i \\ \vdots \\ 2^N-1+(2^N+1)i \end{pmatrix},\tag{4.6}$$

where c_4 is the normalization constant given by,

$$c_4 = \sum_{x=1}^{2^N}((x-1)^2 + (x+1)^2).\tag{4.7}$$

Complex density matrices used for demonstration are,

$$\rho_1 = \begin{pmatrix} 0.4 & -0.2-0.1i \\ -0.2+0.1i & 0.6 \end{pmatrix},\tag{4.8}$$

$$\rho_2 = \begin{pmatrix} 0.4 & -0.1-0.4i \\ -0.1+0.4i & 0.6 \end{pmatrix}.\tag{4.9}$$

All the methods which will be discussed in this chapter are written inside the class **GatesTools**, and this whole class is written inside the Python module **chap4_quantumtools.py**. To import the class **GatesTools** from the **QuantumInformation** library, and creating the object of the class in your Python code can be done as follows,

```
# importing the class GatesTools from the QuantumInformation library
from QuantumInformation import GatesTools as GT

# creating the object of the class
gt_obj=GT()
```

4.1   Frequently used quantum gates

Quantum gates are important elements in a quantum computing algorithm. In a quantum circuit, the quantum logic gate is a rudimentary part of the quantum circuit that performs operations on a finite number of qubits. In contrast to classical logic gates, the quantum logic gates are reversible. Quantum logic gates manifest themselves as unitary matrices and they operate on a quantum state to produce a desired output depending on what operation they do. Gates are classified as one-qubit, two-qubit and multi-qubit depending on the state on which they act. Just as how in Boolean algebra we have truth tables for the operations, similarly we have them in the case of quantum gates too. We will be giving the numerical recipes to construct such gates; however, one can build up complex quantum circuits using these gates as the basic building blocks.

4.1.1 The Pauli-X

The Pauli-X gate acts on a single-qubit state. This is the quantum gate equivalent to the NOT gate. It is sometimes also known as a bit flip operator. It maps $|0\rangle$ to $|1\rangle$ and $|1\rangle$ to $|0\rangle$. A generalized version of writing the Pauli-X acting on a qubit will be $X|a\rangle = |1-a\rangle$, where a is either 1 or 0. The matrix representation of Pauli-X is given below:

$$X = \begin{pmatrix} 0 & 1 \\ 1 & 0 \end{pmatrix}.$$

To flip a N-qubit computational basis, we have to operate N tensor products of the Pauli-X matrix, and the mathematical operation is shown below,

$$X^{\otimes N}|a_1 a_2 \ldots a_N\rangle = |1-a_1\rangle \otimes |1-a_2\rangle \ldots |1-a_N\rangle \tag{4.10}$$

Circuit symbol

Or,

Truth table

Input	Output
$|0\rangle$	$|1\rangle$
$|1\rangle$	$|0\rangle$

`sx(N=1)`

Parameters

in/out	Argument	Description
[in]	N	It signifies the number of qubits. Its default value is equal to 1
[out]	sigmax	sigmax is a real array of dimension $(0{:}2^N{-}1, 0{:}2^N{-}1)$, which is the N-qubit Pauli-X matrix

Implementation

```
def sx(self,N=1):
    """

    Construct N-qubit Pauli spin matrix sigma_x
    Inputs:
        N: number of spins
    Output:
        sigmax: It stores the Pauli spin matrix sx
    """
    sigmax=np.zeros([2**N,2**N])
    j=(2**N)-1
    for i in range(0,2**N):
        sigmax[i,j]=1
        j=j-1
    return sigmax
```

Example

In the following example, we construct the $X^{\otimes 2}$ matrix,

```
B=gt_obj.sx(N=2)
print(B)
```

The above code prints the following output,

```
[[0. 0. 0. 1.]
 [0. 0. 1. 0.]
 [0. 1. 0. 0.]
 [1. 0. 0. 0.]]
```

4.1.2   The Pauli-Y

The Pauli-Y gate acts on a single-qubit state. It maps $|0\rangle$ to $i|1\rangle$ and $|1\rangle$ to $-i|0\rangle$. It is clear that the Pauli-Y gate is similar to the Pauli-X gate but with an additional phase of $\pm i$. A generalized version of writing the Pauli-Y acting on a qubit will be $Y|a\rangle = (-1)^a i|1 - a\rangle$ where a is 0 or 1. The matrix representation of Pauli-Y is given below:

$$Y = \begin{pmatrix} 0 & -i \\ i & 0 \end{pmatrix}.$$

The N-qubit operation of the Pauli-Y matrix on a N-qubit computational basis is shown below,

$$Y^{\otimes N}|a_1 a_2 \ldots a_N\rangle = (-1)^{a_1+a_2+\ldots+a_N}(i^N)|1 - a_1\rangle \otimes |1 - a_2\rangle \otimes \ldots \otimes |1 - a_N\rangle \qquad (4.11)$$

Circuit symbol

Truth table

Input	Output
$|0\rangle$	$i|1\rangle$
$|1\rangle$	$-i|0\rangle$

`sy(N=1)`

Parameters

in/out	Argument	Description
[in]	N	It signifies the number of qubits. Its default value is equal to 1
[out]	sigmay	sigmay is a complex array of dimension $(0{:}2^N-1,0{:}2^N-1)$, which is the N-qubit Pauli-Y matrix

Implementation

```python
def sy(self,N=1):
    """
    Construct N-qubit Pauli spin matrix sigma_y
    Inputs:
        N: Number of spins
    Outputs:
        sigmay: It stores the Pauli spin matrix sy
    """
    sigmay2=np.array([[0,complex(0,-1)],[complex(0,1),0]])
    if N >1:
        for i in range(2,N+1):
            if i==2:
                sigmay=np.kron(sigmay2, sigmay2)
            elif i > 2:
                sigmay=np.kron(sigmay, sigmay2)
    else:
        sigmay=sigmay2

    return sigmay
```

Example

In the following example, we construct the $Y^{\otimes 2}$ matrix.

```python
B=gt_obj.sy(N=2)
print(B)
```

The above code prints the $Y^{\otimes 2}$ matrix.

```
[[ 0.+0.j   0.+0.j   0.+0.j  -1.-0.j]
 [ 0.+0.j   0.+0.j   1.+0.j   0.+0.j]
 [ 0.+0.j   1.+0.j   0.+0.j   0.+0.j]
 [-1.+0.j   0.+0.j   0.+0.j   0.+0.j]]
```

4.1.3   The Pauli-Z

The Pauli-Z gate acts on a single-qubit state. It leaves $|0\rangle$ unchanged and maps $|1\rangle$ to $-|1\rangle$, the reason being, the Pauli-Z gate is already in the diagonal basis. A generalized version of writing the Pauli-Z acting on a qubit will be $Z|a\rangle = (1 - 2a)|a\rangle$, where a is 0 or 1. The matrix representation of Pauli-Z is given below:

$$Z = \begin{pmatrix} 1 & 0 \\ 0 & -1 \end{pmatrix}.$$

The N-qubit operation of the Pauli-Z matrix on a N-qubit computational basis is shown below,

$$Z^{\otimes N}|a_1 a_2 \ldots a_N\rangle = (1 - 2a_1)(1 - 2a_2)\ldots(1 - 2a_N)|a_1 a_2 \ldots a_N\rangle \tag{4.12}$$

Circuit symbol

$$-\boxed{Z}-$$

Truth table

Input	Output
$|0\rangle$	$|0\rangle$
$|1\rangle$	$-|1\rangle$

sz(N=1)

Parameters

in/out	Argument	Description
[in]	N	It signifies the number of qubits. Its default value is equal to 1
[out]	sigmaz	sigmaz is a real array of dimension $(0{:}2^N - 1, 0{:}2^N - 1)$, which is the N-qubit Pauli-Z matrix

Implementation

```
def sz(self,N=1):
    """
    Construct N-qubit Pauli spin matrix sigma_z
    Inputs:
        N: Number of spins
    Outputs:
        sigmaz: It stores the Pauli spin matrix sz
    """
    sigmaz2=np.array([[1,0],[0,-1]])
    if N >1:
        for i in range(2,N+1):
            if i==2:
                sigmaz=np.kron(sigmaz2, sigmaz2)
            elif i > 2:
                sigmaz=np.kron(sigmaz, sigmaz2)
    else:
        sigmaz=sigmaz2

    return sigmaz
```

Example

In the following example, we construct the $Z^{\otimes 2}$ matrix.

```
B=gt_obj.sz(N=2)
print(B)
```

The above code prints the $Z^{\otimes 2}$ matrix.

```
[[ 1  0  0  0]
 [ 0 -1  0  0]
 [ 0  0 -1  0]
 [ 0  0  0  1]]
```

4.1.4   The Hadamard gate

The Hadamard gate acts on a single-qubit state. It maps $|0\rangle$ to $\frac{|0\rangle+|1\rangle}{\sqrt{2}}$ and $|1\rangle$ to $\frac{|0\rangle-|1\rangle}{\sqrt{2}}$. The Hadamard gate is often used to interconvert between the Pauli-X and the Pauli-Z eigenstates. It is to be noted that the columns of the Hadamard matrix are the eigenvectors of Pauli-X when represented in the computational basis. A generalized version of writing the Hadmard gate acting on a qubit will be $H|a\rangle = \frac{|0\rangle + (-1)^a|1\rangle}{\sqrt{2}}$, where a is 0 or 1. The matrix representation of the Hadamard gate is given below:

$$H = \frac{1}{\sqrt{2}} \begin{pmatrix} 1 & 1 \\ 1 & -1 \end{pmatrix}.$$

The N-qubit operation of the Hadamard matrix on a N-qubit computational basis is shown below,

$$H^{\otimes N}|a_1 a_2 \ldots a_N\rangle = \frac{|0\rangle + (-1)^{a_1}|1\rangle}{\sqrt{2}} \otimes \frac{|0\rangle + (-1)^{a_2}|1\rangle}{\sqrt{2}} \otimes \ldots \otimes \frac{|0\rangle + (-1)^{a_N}|1\rangle}{\sqrt{2}} \quad (4.13)$$

The N-qubit Hadamard matrix is quite ubiquitous like for example it is used in Deutsch–Jozsa algorithm, where it transforms the qubits in the query register.

Circuit symbol

Truth table

Input	Output
$|0\rangle$	$\frac{|0\rangle + |1\rangle}{\sqrt{2}}$
$|1\rangle$	$\frac{|0\rangle - |1\rangle}{\sqrt{2}}$

`hadamard_mat(N=1)`

Parameters

in/out	Argument	Description
[in]	N	It signifies the number of qubits. Its default value is equal to 1
[out]	hadamard	It is a real array of dimension $(0:2^N-1, 0:2^N-1)$, which is the N-qubit Hadamard matrix

Implementation

```
def hadamard_mat(self,N=1):
    """
    Construct N-qubit Hadamard matrix
    Inputs:
        N: Number of spins
    Outputs:
        hadamard: It stores the Hadamard matrix
    """
    hadamard2=np.array([[1/np.sqrt(2),1/np.sqrt(2)],\
                        [1/np.sqrt(2),-1/np.sqrt(2)]])
    if N >1:
        for i in range(2,N+1):
            if i==2:
                hadamard=np.kron(hadamard2, hadamard2)
            elif i > 2:
```

```
                hadamard=np.kron(hadamard, hadamard2)
    else:
        hadamard=hadamard2

    return hadamard
```

Example

In the following example, we construct the $H^{\otimes 2}$ matrix,

```
B=gt_obj.hadamard_mat(N=2)
print(B)
```

The above code prints the $H^{\otimes 2}$ matrix.

```
[[ 0.5  0.5  0.5  0.5]
 [ 0.5 -0.5  0.5 -0.5]
 [ 0.5  0.5 -0.5 -0.5]
 [ 0.5 -0.5 -0.5  0.5]]
```

4.1.5 The phase gate

The phase gate (S) acts on a single-qubit state. It leaves $|0\rangle$ unchanged and maps $|1\rangle$ to $i|1\rangle$, where i is a phase factor. A generalized version of writing the phase gate acting on a qubit will be $S|a\rangle = i^a |a\rangle$, where a is 0 or 1. The matrix representation of the phase gate is given below:

$$S = \begin{pmatrix} 1 & 0 \\ 0 & i \end{pmatrix}.$$

The N-qubit operation of the phase gate on a N-qubit computational basis is shown below,

$$S^{\otimes N}|a_1 a_2 \ldots a_N\rangle = i^{a_1 + a_2 + \ldots + a_N}|a_1 a_2 \ldots a_N\rangle \tag{4.14}$$

Circuit symbol

Truth table

Input	Output
$|0\rangle$	$|0\rangle$
$|1\rangle$	$i|1\rangle$

```
phase_gate(N=1)
```

Parameters

in/out	Argument	Description
[in]	N	It signifies the number of qubits. Its default value is equal to 1
[out]	phaseg	It is a complex array of dimension $(0{:}2^N-1,0{:}2^N-1)$, which is the N-qubit phase gate

Implementation

```
def phase_gate(self,N=1):
    """

    Construct N-qubit phase gate matrix
    Inputs:
        N: Number of spins
    Outputs:
        phaseg: It stores the phase gate matrix
    """
    phaseg2=np.array([[1,0],\
                      [0,complex(0,1)]])
    if N >1:
        for i in range(2,N+1):
            if i==2:
                phaseg=np.kron(phaseg2, phaseg2)
            elif i > 2:
                phaseg=np.kron(phaseg, phaseg2)
    else:
        phaseg=phaseg2

    return phaseg
```

Example

In the following example, we construct the $S^{\otimes 2}$ matrix,

```
B=gt_obj.phase_gate(N=2)
print(B)
```

The above code prints the $S^{\otimes 2}$ matrix.

```
[[ 1.+0.j  0.+0.j  0.+0.j  0.+0.j]
 [ 0.+0.j  0.+1.j  0.+0.j  0.+0.j]
 [ 0.+0.j  0.+0.j  0.+1.j  0.+0.j]
 [ 0.+0.j  0.+0.j  0.+0.j -1.+0.j]]
```

4.1.6   The rotation gate

The rotation gate R_k acts on a single-qubit state. It leaves $|0\rangle$ unchanged and maps $|1\rangle$ to $e^{\frac{2\pi i}{2^k}}|1\rangle$, where k is a positive number. A generalized version of writing the rotation gate acting on a qubit will be $R_k|a\rangle = e^{\frac{2\pi i}{2^k} \times a}|a\rangle$, where a is 0 or 1. The matrix representation of the rotation gate is given below:

$$R_k = \begin{pmatrix} 1 & 0 \\ 0 & e^{\frac{2\pi i}{2^k}} \end{pmatrix}.$$

The N-qubit operation of the rotation matrix on a N-qubit computational basis is shown below,

$$R_k^{\otimes N}|a_1 a_2 \ldots a_N\rangle = e^{\frac{2\pi i}{2^k} \times a_1} e^{\frac{2\pi i}{2^k} \times a_2} \ldots e^{\frac{2\pi i}{2^k} \times a_N}|a_1 a_2 \ldots a_N\rangle \quad (4.15)$$

Circuit symbol

Truth table

Input	Output
$|0\rangle$	$|0\rangle$
$|1\rangle$	$e^{\frac{2\pi i}{2^k}}|1\rangle$

Note that when $k = 2$, we get the phase gate S, meaning that $S = R_2$.

`rotation_gate(k,N=1)`

Parameters

in/out	Argument	Description
[in]	k	k is a positive integer
[in]	N	It signifies the number of qubits. Its default value is equal to 1
[out]	rotg	It is a complex array of dimension $(0:2^N-1,0:2^N-1)$, which is the N-qubit rotation gate

Implementation

```
def rotation_gate(self,k,N=1):
    """
    Construct N-qubit rotation gate matrix
    Input:
        k: is a positive number
        N: number of spins
    Returns:
        rotg: Rotation gate matrix
    """
    assert k > 0, "k is not positive number"
```

```
z=complex(0,(2*math.pi)/(2**k))
rotg2=np.array([[1,0],[0,cmath.exp(z)]])
if N >1:
    for i in range(2,N+1):
        if i==2:
            rotg=np.kron(rotg2, rotg2)
        elif i > 2:
            rotg=np.kron(rotg, rotg2)
else:
    rotg=rotg2

return rotg
```

Example

In the following example, we construct the $R_3^{\otimes 2}$ matrix,

```
B=gt_obj.rotation_gate(k=3, N=2)
print(B)
```

The above code prints the $R_3^{\otimes 2}$ matrix.

```
[[1.00000000e+00+0.j    0.00000000e+00+0.j    0.00000000e+00+0.j
   0.00000000e+00+0.j]
 [0.00000000e+00+0.j    7.07106781e-01+0.70710678j    0.00000000e+00+0.j
   0.00000000e+00+0.j]
 [0.00000000e+00+0.j    0.00000000e+00+0.j    7.07106781e-01+0.70710678j
   0.00000000e+00+0.j]
 [0.00000000e+00+0.j    0.00000000e+00+0.j    0.00000000e+00+0.j
   2.22044605e-16+1.j]]
```

4.1.7 The controlled NOT gate (CX gate)

The controlled NOT gate acts on a two-qubit state. The controlled-NOT X gate performs a Pauli-X operation on the second qubit (called target bit) only when the first qubit (called control bit) is $|1\rangle$, otherwise leaves it unchanged. With respect to the standard two-qubit orthonormal basis $\{|00\rangle, |01\rangle, |10\rangle, |11\rangle\}$. The matrix representation of the controlled NOT gate is given below:

$$CX = \begin{pmatrix} 1 & 0 & 0 & 0 \\ 0 & 1 & 0 & 0 \\ 0 & 0 & 0 & 1 \\ 0 & 0 & 1 & 0 \end{pmatrix}.$$

Circuit symbol

Or,

Truth table

Input	Output
$|00\rangle$	$|00\rangle$
$|01\rangle$	$|01\rangle$
$|10\rangle$	$|11\rangle$
$|11\rangle$	$|10\rangle$

`cx_gate()`

Parameters

in/out	Argument	Description
[out]	A	It is a real array of dimension (0:3,0:3), which is the controlled NOT gate

Implementation

```
def cx_gate(self):
    """
    It returns controlled NOT gate
    """
    return np.array([[1,0,0,0],[0,1,0,0],[0,0,0,1],[0,0,1,0]])
```

Example

In this example, we are illustrating the fourth operation of the truth table which is $CX|11\rangle = |10\rangle$.

```
A=gt_obj.cx_gate()
state=np.array([0,0,0,1])
res_state=np.matmul(A,state)
print(res_state)
```

The above code prints the output of the fourth operation of the truth table.

```
[0 0 1 0]
```

4.1.8 The controlled Z gate (CZ gate)

The controlled Z gate acts on a two-qubit state. The controlled NOT Z gate performs a Pauli-Z operation on the second qubit (target bit) only when the first qubit (control bit) is $|1\rangle$. The matrix representation of the controlled Z gate is given below:

$$CZ = \begin{pmatrix} 1 & 0 & 0 & 0 \\ 0 & 1 & 0 & 0 \\ 0 & 0 & 1 & 0 \\ 0 & 0 & 0 & -1 \end{pmatrix}.$$

Circuit symbol

Truth table

Input	Output
$|00\rangle$	$|00\rangle$
$|01\rangle$	$|01\rangle$
$|10\rangle$	$|10\rangle$
$|11\rangle$	$-|11\rangle$

`cz_gate()`

Parameters

in/out	Argument	Description
[out]	A	A is a real array of dimension (0:3,0:3), which is the controlled Z gate

Implementation

```
def cz_gate(self):
    """
    It returns controlled Z gate
    """
    return np.array([[1,0,0,0],[0,1,0,0],[0,0,1,0],[0,0,0,-1]])
```

Example

In this example, we are illustrating the fourth operation of the truth table which is $CZ|11\rangle = -|11\rangle$.

```
A=gt_obj.cz_gate()
state=np.array([0,0,0,1])
res_state=np.matmul(A,state)
print(res_state)
```

The above code prints the output of the fourth operation of the truth table.

```
[ 0  0  0 -1]
```

4.1.9 The SWAP gate

The SWAP gate swaps or permutes the bits of a two-qubit state. For example, the SWAP gate operating on $|10\rangle$ will result in $|01\rangle$. The matrix representation of the SWAP gate is given below:

$$SWAP = \begin{pmatrix} 1 & 0 & 0 & 0 \\ 0 & 0 & 1 & 0 \\ 0 & 1 & 0 & 0 \\ 0 & 0 & 0 & 1 \end{pmatrix}.$$

Circuit symbol

Truth table

Input	Output
$|00\rangle$	$|00\rangle$
$|01\rangle$	$|10\rangle$
$|10\rangle$	$|01\rangle$
$|11\rangle$	$|11\rangle$

```
swap_gate()
```

Parameters

in/out	Argument	Description
[out]	A	A is a real array of dimension (0:3,0:3), which is the SWAP gate

Implementation

```
def swap_gate(self):
    """
    It returns a swap gate
    """
    return np.array([[1,0,0,0],[0,0,1,0],[0,1,0,0],[0,0,0,1]])
```

Example

In this example, we are illustrating the third operation of the truth table which is SWAP$|10\rangle = |01\rangle$.

```
A=gt_obj.swap_gate()
state=np.array([0,0,1,0])
res_state=np.matmul(A,state)
print(res_state)
```

The above code prints the output of the third operation of the truth table.

```
[0 1 0 0]
```

4.1.10   The Toffoli gate (CCX, CCNOT, TOFF)

The Toffoli gate [42, 43], named after Tommaso Toffoli; is also called the CCNOT gate. The Toffoli gate acts on a three-qubit state. When the first two qubits are $|1\rangle$ then only it performs a Pauli-X operation on the third qubit. The matrix representation of the Toffoli gate is given below:

$$
CCX = \begin{pmatrix}
1 & 0 & 0 & 0 & 0 & 0 & 0 & 0 \\
0 & 1 & 0 & 0 & 0 & 0 & 0 & 0 \\
0 & 0 & 1 & 0 & 0 & 0 & 0 & 0 \\
0 & 0 & 0 & 1 & 0 & 0 & 0 & 0 \\
0 & 0 & 0 & 0 & 1 & 0 & 0 & 0 \\
0 & 0 & 0 & 0 & 0 & 1 & 0 & 0 \\
0 & 0 & 0 & 0 & 0 & 0 & 0 & 1 \\
0 & 0 & 0 & 0 & 0 & 0 & 1 & 0
\end{pmatrix}.
$$

Circuit symbol

Truth table

Input	Output
$|000\rangle$	$|000\rangle$
$|001\rangle$	$|001\rangle$
$|010\rangle$	$|010\rangle$
$|011\rangle$	$|011\rangle$
$|100\rangle$	$|100\rangle$
$|101\rangle$	$|101\rangle$
$|110\rangle$	$|111\rangle$
$|111\rangle$	$|110\rangle$

`toffoli_gate()`

Parameters

in/out	Argument	Description
[out]	A	A is a real array of dimension (0:7,0:7), which is the Toffoli gate

Implementation

```
def toffoli_gate(self):
    """
    It returns a Toffoli gate
    """
    return np.array([[1,0,0,0,0,0,0,0],[0,1,0,0,0,0,0,0],\
                     [0,0,1,0,0,0,0,0],[0,0,0,1,0,0,0,0],\
                     [0,0,0,0,1,0,0,0],[0,0,0,0,0,1,0,0],\
                     [0,0,0,0,0,0,0,1],[0,0,0,0,0,0,1,0]])
```

Example

In this example, we are illustrating the seventh operation of the truth table which is $CCX|110\rangle = |111\rangle$.

```
A=gt_obj.toffoli_gate()
state=np.array([0,0,0,0,0,0,1,0])
res_state=np.matmul(A,state)
print(res_state)
```

The above code prints the output of the seventh operation of the truth table.

```
[0 0 0 0 0 0 0 1]
```

4.1.11   The Fredkin gate (CSWAP)

The Fredkin gate [44] (also CSWAP or cS gate), named after Edward Fredkin, is a three qubit gate that performs a controlled swap operation. It simply means that, when the control bit is 1, it swaps the second and the third bit. An interesting feature of the Fredkin gate is that it can be used as a universal gate and also to do reversible computing. The matrix representation of the Fredkin gate is given below:

$$F = \begin{pmatrix} 1 & 0 & 0 & 0 & 0 & 0 & 0 & 0 \\ 0 & 1 & 0 & 0 & 0 & 0 & 0 & 0 \\ 0 & 0 & 1 & 0 & 0 & 0 & 0 & 0 \\ 0 & 0 & 0 & 1 & 0 & 0 & 0 & 0 \\ 0 & 0 & 0 & 0 & 1 & 0 & 0 & 0 \\ 0 & 0 & 0 & 0 & 0 & 0 & 1 & 0 \\ 0 & 0 & 0 & 0 & 0 & 1 & 0 & 0 \\ 0 & 0 & 0 & 0 & 0 & 0 & 0 & 1 \end{pmatrix}.$$

Truth table

Input	Output
$|000\rangle$	$|000\rangle$
$|001\rangle$	$|001\rangle$
$|010\rangle$	$|010\rangle$
$|011\rangle$	$|011\rangle$
$|100\rangle$	$|100\rangle$
$|101\rangle$	$|110\rangle$
$|110\rangle$	$|101\rangle$
$|111\rangle$	$|111\rangle$

`fredkin_gate()`

Parameters

in/out	Argument	Description
[out]	A	A is a real array of dimension (0:7,0:7), which is the Fredkin gate

Implementation

```
def fredkin_gate(self):
    """
    It returns a Fredkin gate
    """
    return np.array([[1,0,0,0,0,0,0,0],[0,1,0,0,0,0,0,0],\
                     [0,0,1,0,0,0,0,0],[0,0,0,1,0,0,0,0],\
                     [0,0,0,0,1,0,0,0],[0,0,0,0,0,0,1,0],\
                     [0,0,0,0,0,1,0,0],[0,0,0,0,0,0,0,1]])
```

Example

In this example, we are illustrating the seventh operation of the truth table which is $F|110\rangle = |101\rangle$.

```
A=gt_obj.fredkin_gate()
state=np.array([0,0,0,0,0,0,1,0])
res_state=np.matmul(A,state)
print(res_state)
```

The above code prints the output of the seventh operation of the truth table.

```
[0 0 0 0 0 1 0 0]
```

4.2   The Bell states

The Bell states are four specific maximally entangled two-qubit states [24, 45] as shown below. It is also worth to mention that they contribute another set of orthonormal basis in the two-qubit Hilbert space apart from the standard two-qubit basis. Bell states are very important from the point of view of operations such as teleportation [46], superdense coding [47] and to produce a simplified version of the EPR paradox [48].

$$|b_1\rangle = \frac{1}{\sqrt{2}}(|01\rangle + |10\rangle), \tag{4.16}$$

$$|b_2\rangle = \frac{1}{\sqrt{2}}(|01\rangle - |10\rangle), \tag{4.17}$$

$$|b_3\rangle = \frac{1}{\sqrt{2}}(|00\rangle + |11\rangle), \tag{4.18}$$

$$|b_4\rangle = \frac{1}{\sqrt{2}}(|00\rangle - |11\rangle). \tag{4.19}$$

The states $|b_1\rangle$, $|b_2\rangle$, $|b_3\rangle$ and $|b_4\rangle$ together form a maximally entangled two-qubit basis of the four-dimensional Hilbert space called the Bell basis. Bell states have many applications in the field of quantum information, quantum computation and primarily in the quantum spin chain system. For example, in the Majumdar-Ghosh model, the ground states of the model at the MG point are doubly degenerate, and they are the tensor product of $|b_2\rangle$ state.

The doubly degenerate ground states can be simply written as,

$$|R_N\rangle = [1\ 2] \otimes [3\ 4] \otimes \ldots \otimes [N-1\ N], \qquad |L_N\rangle = [2\ 3] \otimes [4\ 5] \otimes \ldots \otimes [N\ 1], \qquad (4.20)$$

where each of the two-qubit elemental blocks are, $[i\ i+1] = \dfrac{1}{\sqrt{2}}(|01\rangle_{ii+1} - |10\rangle_{ii+1})$, and N (even number) represents the system size. From Eq. [4.20] we observe that the $|L_N\rangle$ state can be constructed by translationally shifting the $|R_N\rangle$ state to the right. Based on the above discussion, in our following methods, we provide the options of constructing N tensor product of any of the four Bell states and also its translationally shifted state.

4.2.1 $N/2$ tensor product of $|b_1\rangle$ Bell state

In this method, we can construct any one of the two states,

$$|\phi_1\rangle = \frac{1}{2^{N/4}}\left(|01\rangle_{12} + |10\rangle_{12}\right) \otimes \left(|01\rangle_{34} + |10\rangle_{34}\right) \otimes \ldots \otimes \left(|01\rangle_{N-1N} + |10\rangle_{N-1N}\right), \quad (4.21)$$

and

$$|\phi_2\rangle = \frac{1}{2^{N/4}}\left(|01\rangle_{23} + |10\rangle_{23}\right) \otimes \left(|01\rangle_{45} + |10\rangle_{45}\right) \otimes \ldots \otimes \left(|01\rangle_{N1} + |10\rangle_{N1}\right), \qquad (4.22)$$

The following method uses the function **RecurChainRL1** from the module **RecurNum**, and the functionality of **RecurChainRL1** has been explained at the end of this section.

`bell1(tot_spins=2,shift=0)`

Parameters

in/out	Argument	Description
[in]	tot_spins	It stores the total number of spins (N), and it must be even number. The default value is 2
[in]	shift	It can take only the following values, • shift=0, generates the state $|\phi_1\rangle$. It is the default value • shift=1, generates the state $|\phi_2\rangle$
[out]	state	It is a real array which stores either the state $|\phi_1\rangle$ or $|\phi_2\rangle$.

Implementation

```
def bell1(self,tot_spins=2,shift=0):
    """

    Construct N tensor products of the |bell1> or T|bell1> Bell state
    Input:
        tot_spins: The total number of spins
        shift: for value 0 we get |bell1> and for value 1 T|bell1>.
    Output:
        state: the result will be |bell1> or T|bell1> state.
```

```
"""
# Checking the number of qubits N to be even
assert tot_spins%2==0,"the total number of spins is not an even number"
# Checking the entered shift value is correct or not
assert shift==0 or shift==1, "Invalid entry of the shift value"
terms=int(tot_spins/2)
row=np.zeros([terms,1])
mylist=[]
icount=-1
RecurNum.RecurChainRL1(row,tot_spins,icount,mylist,shift)
mylist=np.array(mylist)
state=np.zeros([2**tot_spins])
factor=1/math.sqrt(2)
len_mylist=len(mylist)
# Constructing the state |φ₁⟩ or |φ₂⟩
for x in range(0,len_mylist):
    state[mylist.item(x)]=factor**terms
return(state)
```

Example

In this example, we are generating the state $|\phi_2\rangle$ of Eq. [4.22], for $N = 4$ total spins.

$$|\phi_2\rangle \quad = \quad \frac{1}{2}(|0101\rangle_{2341} + |0110\rangle_{2341} + |1001\rangle_{2341} + |1010\rangle_{2341}) \qquad (4.23)$$

$$|\phi_2\rangle \quad = \quad \frac{1}{2}(|1010\rangle_{1234} + |0011\rangle_{1234} + |1100\rangle_{1234} + |0101\rangle_{1234}) \qquad (4.24)$$

The matrix representation of the state $|\phi_2\rangle$ in Eq. [4.24] can written as,

$$|\phi_2\rangle = [\,0\ 0\ 0\ 0.5\ 0\ 0.5\ 0\ 0\ 0\ 0\ 0.5\ 0\ 0.5\ 0\ 0\ 0\,]^{1/T} \qquad (4.25)$$

```
state=gt_obj.bell1(4,shift=1)
print(state)
```

The above code prints the following array,

```
[0.  0.  0.  0.5  0.  0.5  0.  0.  0.  0.  0.5  0.  0.5  0.  0.  0.]
```

As stated earlier, we have used the recursive function **RecurChainRL1** in our present method. For understanding purpose of this recursive function, let us consider the example given in Eq. [4.24]. The state $|\phi_2\rangle$ of Eq. [4.24] has non-zero entry at the following places $(10)_{10} \equiv (1010)_2$, $(3)_{10} \equiv (0011)_2$, $(12)_{10} \equiv (1100)_2$, and $(5)_{10} \equiv (0101)_2$. These places are calculated by the recursive function **RecurChainRL1**, and output will be stored in a list as, $[10, 3, 12, 5]$.

4.2.2 $N/2$ tensor product of $|b_2\rangle$ Bell state

In this method, we can construct any one of the two states,

$$|\phi_3\rangle = \frac{1}{2^{N/2}} \left(|01\rangle_{12} - |10\rangle_{12}\right) \otimes \left(|01\rangle_{34} - |10\rangle_{34}\right) \otimes \ldots \otimes \left(|01\rangle_{N-1N} - |10\rangle_{N-1N}\right), \qquad (4.26)$$

and

$$|\phi_4\rangle = \frac{1}{2^{N/2}} \left(|01\rangle_{23} - |10\rangle_{23} \right) \otimes \left(|01\rangle_{45} - |10\rangle_{45} \right) \otimes \ldots \otimes \left(|01\rangle_{N1} - |10\rangle_{N1} \right), \qquad (4.27)$$

The following method uses the function **RecurChainRL2** from the module **RecurNum**, and the functionality of **RecurChainRL2** explained at the end of this section.

`bell2(tot_spins=2,shift=0)`

Parameters

in/out	Argument	Description
[in]	tot_spins	It stores the total number of spins (N), and it must be even number. The default value is 2
[in]	shift	It can take only the following values, • shift=0, generates the state $|\phi_3\rangle$. It is the default value • shift=1, generates the state $|\phi_4\rangle$
[out]	state	It is a real array which stores either the state $|\phi_3\rangle$ or $|\phi_4\rangle$

Implementation

```
def bell2(self,tot_spins=2,shift=0):
    """

    Construct N tensor products of the |bell2> or T|bell2> Bell state
    Input:
        tot_spins: The total number of spins
        shift: for value 0 we get |bell2> and for value 1 T|bell2>.
    Output:
        state: the result will be |bell2> or T|bell2> state.
    """
    # Checking the number of qubits N to be even
    assert tot_spins%2==0,"the total number of spins is not an even number"
    # Checking the entered shift value is correct or not
    assert shift==0 or shift==1, "Invalid entry of the shift value"
    terms=int(tot_spins/2)
    row=np.zeros([terms,1])
    mylist=[]
    icount=-1
    RecurNum.RecurChainRL2(row,tot_spins,icount,mylist,shift)
    mylist=np.array(mylist)
    state=np.zeros([2**tot_spins])
    factor=1/math.sqrt(2)
    len_mylist=len(mylist)
    # Constructing the state |φ₃⟩ or |φ₄⟩
    for x in range(0,len_mylist):
        if mylist.item(x)<0:
```

```
          state[-mylist.item(x)]=-factor**terms
     elif mylist.item(x)>=0:
          state[mylist.item(x)]=factor**terms
   return(state)
```

Example

In this example, we are generating the state $|\phi_4\rangle$ of Eq. [4.27], for $N = 4$ total spins.

$$|\phi_4\rangle = \frac{1}{2}(|0101\rangle_{2341} - |0110\rangle_{2341} - |1001\rangle_{2341} + |1010\rangle_{2341}) \qquad (4.28)$$

$$|\phi_4\rangle = \frac{1}{2}(|1010\rangle_{1234} - |0011\rangle_{1234} - |1100\rangle_{1234} + |0101\rangle_{1234}) \qquad (4.29)$$

The matrix representation of the state $|\phi_4\rangle$ in Eq. [4.29] can written as,

$$|\phi_4\rangle = [\,0\ 0\ 0\ -0.5\ 0\ 0.5\ 0\ 0\ 0\ 0\ 0.5\ 0\ -0.5\ 0\ 0\ 0\,]^{1/T} \qquad (4.30)$$

```
state=gt_obj.bell2(4,shift=1)
print(state)
```

The above code prints the following array,

```
[0.  0.  0.  -0.5  0.  0.5  0.  0.  0.  0.  0.5  0.  -0.5  0.  0.  0.]
```

As stated earlier, we have used the recursive function **RecurChainRL2** in our present method. For understanding purpose of this recursive function, let us consider the example given in Eq. [4.29]. The state $|\phi_4\rangle$ of Eq. [4.29] has non-zero entry at the following places $(10)_{10} \equiv (1010)_2$, $(3)_{10} \equiv (0011)_2$, $(12)_{10} \equiv (1100)_2$, and $(5)_{10} \equiv (0101)_2$. These places are calculated by the recursive function **RecurChainRL2**, and these place values will be stored in list. Additionally a factor of $+1$ or -1 will be multiplied to these place values depending on whether the corresponding coefficient sign is positive or negative, respectively. For the case under consideration, the list which will be generated is $[10, -3, -12, 5]$.

4.2.3 $N/2$ tensor product of $|b_3\rangle$ Bell state

In this method, we can construct any one of the two states,

$$|\phi_5\rangle = \frac{1}{2^{N/2}} \left(|00\rangle_{12} + |11\rangle_{12}\right) \otimes \left(|00\rangle_{34} + |11\rangle_{34}\right) \otimes \ldots \otimes \left(|00\rangle_{N-1N} + |11\rangle_{N-1N}\right), \quad (4.31)$$

and

$$|\phi_6\rangle = \frac{1}{2^{N/2}} \left(|00\rangle_{23} + |11\rangle_{23}\right) \otimes \left(|00\rangle_{45} + |11\rangle_{45}\right) \otimes \ldots \otimes \left(|00\rangle_{N1} + |11\rangle_{N1}\right), \qquad (4.32)$$

The following method uses the function **RecurChainRL3** from the module **RecurNum**, and the functionality of **RecurChainRL3** explained at the end of this section.

```
bell3(tot_spins=2,shift=0)
```

Parameters

in/out	Argument	Description
[in]	tot_spins	It stores the total number of spins (N), and it must be even number. The default value is 2
[in]	shift	It can take only the following values, • shift=0, generates the state $|\phi_5\rangle$. It is the default value • shift=1, generates the state $|\phi_6\rangle$
[out]	state	It is a real array which stores either the state $|\phi_5\rangle$ or $|\phi_6\rangle$

Implementation

```
def bell3(self,tot_spins=2,shift=0):
    """
    Construct N tensor products of the |bell3> or T|bell3> Bell state
    Input:
        tot_spins: The total number of spins
        shift: for value 0 we get |bell3> and for value 1 T|bell3>.
    Output:
        state: the result will be |bell3> or T|bell3> state.
    """
    # Checking the number of qubits N to be even
    assert tot_spins%2==0,"the total number of spins is not an even number"
    # Checking the entered shift value is correct or not
    assert shift==0 or shift==1, "Invalid entry of the shift value"
    terms=int(tot_spins/2)
    row=np.zeros([terms])
    mylist=[]
    icount=-1
    RecurNum.RecurChainRL3(row,tot_spins,icount,mylist,shift)
    mylist=np.array(mylist)
    state=np.zeros([2**tot_spins])
    factor=1/math.sqrt(2)
    len_mylist=len(mylist)
    # Constructing the state |φ5> or |φ6>
    for x in range(0,len_mylist):
        state[mylist.item(x)]=factor**terms
    return(state)
```

Example

In this example, we are generating the state $|\phi_5\rangle$ of Eq. [4.31], for $N = 4$ total spins.

$$|\phi_5\rangle \;\; = \;\; \frac{1}{2}(|0000\rangle_{1234} + |0011\rangle_{1234} + |1100\rangle_{1234} + |1111\rangle_{1234}) \tag{4.33}$$

The matrix representation of the state $|\phi_5\rangle$ in Eq. [4.33] can written as,

$$|\phi_5\rangle = \left[\, 0.5\ 0\ 0\ 0\ 0.5\ 0\ 0\ 0\ 0\ 0\ 0\ 0\ 0.5\ 0\ 0\ 0.5 \,\right]^{1/T} \tag{4.34}$$

```
state=gt_obj.bell3(4,shift=0)
print(state)
```

The above code prints the following array,

```
[0.5  0.   0.   0.5  0.   0.   0.   0.   0.   0.   0.   0.   0.5  0.   0.   0.5]
```

As stated earlier, we have used the recursive function **RecurChainRL3** in our present method. For understanding purpose of this recursive function, let us consider the example given in Eq. [4.33]. The state $|\phi_5\rangle$ of Eq. [4.33] has non-zero entry at the following places $(0)_{10} \equiv (0000)_2$, $(3)_{10} \equiv (0011)_2$, $(12)_{10} \equiv (1100)_2$, and $(15)_{10} \equiv (1111)_2$. These places are calculated by the recursive function **RecurChainRL3**, and output will be stored in a list as, $[0, 3, 12, 15]$.

4.2.4 $N/2$ tensor product of $|b_4\rangle$ Bell state

In this method, we can construct any one of the two states,

$$|\phi_7\rangle = \frac{1}{2^{N/2}}\left(|00\rangle_{12} - |11\rangle_{12}\right) \otimes \left(|00\rangle_{34} - |11\rangle_{34}\right) \otimes \ldots \otimes \left(|00\rangle_{N-1N} - |11\rangle_{N-1N}\right), \tag{4.35}$$

and

$$|\phi_8\rangle = \frac{1}{2^{N/2}}\left(|00\rangle_{23} - |11\rangle_{23}\right) \otimes \left(|00\rangle_{45} - |11\rangle_{45}\right) \otimes \ldots \otimes \left(|00\rangle_{N1} - |11\rangle_{N1}\right), \tag{4.36}$$

The following method uses the function **RecurChainRL4** from the module **RecurNum**, and the functionality of **RecurChainRL4** explained at the end of this section.

```
bell3(tot_spins=2,shift=0)
```

Parameters

in/out	Argument	Description
[in]	tot_spins	It stores the total number of spins (N), and it must be even number. The default value is 2
[in]	shift	It can take only the following values, • shift=0, generates the state $|\phi_7\rangle$. It is the default value • shift=1, generates the state $|\phi_8\rangle$
[out]	state	It is a real array which stores either the state $|\phi_7\rangle$ or $|\phi_8\rangle$

Implementation

```python
def bell4(self,tot_spins=2,shift=0):
    """
    Construct N tensor products of the |bell4> or T|bell4> Bell state
    Input:
        tot_spins: The total number of spins
        shift: for value 0 we get |bell4> and for value 1 T|bell4>.
    Output:
        state: the result will be |bell4> or T|bell4> state.
    """
    # Checking the number of qubits N to be even
    assert tot_spins%2==0,"the total number of spins is not an even number"
    # Checking the entered shift value is correct or not
    assert shift==0 or shift==1, "Invalid entry of the shift value"
    terms=int(tot_spins/2)
    row=np.zeros([terms,1])
    mylist=[]
    icount=-1
    RecurNum.RecurChainRL4(row,tot_spins,icount,mylist,shift)
    mylist=np.array(mylist)
    state=np.zeros([2**tot_spins])
    factor=1/math.sqrt(2)
    len_mylist=len(mylist)
    # Constructing the state |φ₇> and |φ₈>
    for x in range(0,len_mylist):
        if mylist.item(x)<0:
            state[-mylist.item(x)]=-factor**terms
        elif mylist.item(x)>=0:
            state[mylist.item(x)]=factor**terms
    return(state)
```

Example

In this example, we are generating the state $|\phi_7\rangle$ of Eq. [4.35], for $N = 4$ total spins.

$$|\phi_7\rangle = \frac{1}{2}(|0000\rangle_{1234} - |0011\rangle_{1234} - |1100\rangle_{1234} + |1111\rangle_{1234}) \tag{4.37}$$

The matrix representation of the state $|\phi_7\rangle$ in Eq. [4.37] can written as,

$$|\phi_7\rangle = \begin{bmatrix} 0.5 \ 0 \ 0 \ -0.5 \ 0 \ 0 \ 0 \ 0 \ 0 \ 0 \ 0 \ 0 \ -0.5 \ 0 \ 0 \ 0.5 \end{bmatrix}^{1/T} \tag{4.38}$$

```python
state=gt_obj.bell4(4,shift=0)
print(state)
```

The above code prints the following array,

```
[0.5  0.   0.   -0.5  0.   0.   0.   0.   0.   0.   0.   0.   -0.5  0.   0.   0.5]
```

As stated earlier, we have used the recursive function **RecurChainRL4** in our present method. For understanding purpose of this recursive function, let us consider the example given in Eq. [4.37]. The state $|\phi_7\rangle$ of Eq. [4.37] has non-zero entry at the following places $(10)_{10} \equiv (1010)_2$, $(3)_{10} \equiv (0011)_2$, $(12)_{10} \equiv (1100)_2$, and $(5)_{10} \equiv (0101)_2$. These places are calculated by the recursive function **RecurChainRL4**, and these place values will be stored in list. Additionally, a factor of $+1$ or -1 will be multiplied to this place values depending on whether the corresponding coefficient sign is positive or negative, respectively. For the case under consideration, the list which will be generated is $[0, -3, -12, 15]$.

4.3   The N-qubit Greenberger-Horne-Zeilinger (GHZ) state

The GHZ state was first studied by Daniel Greenberger, Michael Horne and Anton Zeilinger [49]. The GHZ state is an orthonormal entangled state of three qubits which is,

$$|GHZ\rangle = \frac{|000\rangle + |1111\rangle}{\sqrt{2}}. \tag{4.39}$$

A generalization of it to N qubits ($N \geq 3$) gives us the N-qubit GHZ state which can be defined as

$$|GHZ\rangle = \frac{|0\rangle^{\otimes N} + |1\rangle^{\otimes N}}{\sqrt{2}} = \frac{|00000\cdots0\rangle_N + |11111\cdots1\rangle_N}{\sqrt{2}}. \tag{4.40}$$

These states are extensively used in quantum teleportation [50, 51], quantum cryptography protocols [52–54] and quantum game theory [55]. It is in fact quite interesting to note that the N-qubit GHZ state is the ground state of a ferromagnet [56]. Another interesting property of the three-qubit GHZ state ($N = 3$ case) is that all the three qubits are entangled but no two-qubits are entangled.

```
nGHZ(tot_spins=3)
```

Parameters

in/out	Argument	Description
[in]	tot_spins	It is the total number of qubits. Its default value is equal to 3
[out]	state	It is a real array which stores the GHZ state

Implementation

```
def nGHZ(self,tot_spins=3):
    """
    Construct N-qubit GHZ state
    Input:
        tot_spins: it is the total number of spins, it should be equal to
                   or greater than 3.
    Output:
        state: N-qubit GHZ state.
    """
```

```
# Checking number of spins N ≥ 3
assert tot_spins >= 3, "Total number of spins are less than 3"
state=np.zeros([2**tot_spins])
# Constructing the |GHZ⟩ state
state[0]=1/np.sqrt(2)
state[(2**tot_spins)-1]=1/np.sqrt(2)
return state
```

Example

```
state=gt_obj.nGHZ(tot_spins=3)
print(state)
```

The above code prints the three qubit GHZ state.

```
[0.70710678 0.         0.         0.         0.         0.         0.
          0.70710678]
```

4.4   The N-qubit W state

The W state [57] is an entangled quantum state of three qubits which can be written as

$$|W\rangle = \frac{1}{\sqrt{3}}\left(|100\rangle + |010\rangle + |001\rangle\right). \tag{4.41}$$

It is very different from the three qubit GHZ state in terms of entanglement properties, that is, all the three qubits are entangled and even the two-qubit subsystems are entangled. It is also interesting to note that the three states making up the W state are such that there is a shift of the qubit $|1\rangle$, either to the right or left with respect to the first state. A natural generalization of the W state to the N qubit case is as follows,

$$|W\rangle = \frac{1}{\sqrt{N}}\left(|100\cdots0\rangle + |010\cdots0\rangle + \cdots + |000\cdots1\rangle\right). \tag{4.42}$$

```
nW(tot_spins=3)
```

Parameters

in/out	Argument	Description
[in]	tot_spins	It is the number of qubits. Its default value is equal to 3
[out]	state	It is a real array which stores the W state

Implementation

```
def nW(self,tot_spins=3):
    """
    Construct N-qubit W state
    Input:
        tot_spins: it is the total number of spins, it should be equal to
                   or greater than 3.
    Output:
        state: N-qubit W state.
    """
    # Checking number of spins N ≥ 3
    assert tot_spins >= 3, "Total number of spins are less than 3"
    # Constructing the |W⟩ state
    state=np.zeros([2**tot_spins])
    for i in range(0,tot_spins):
        state[2**i]=1/np.sqrt(tot_spins)
    return state
```

Example

```
state=gt_obj.nW(tot_spins=3)
print(state)
```

The above code prints the three-qubit W state.

```
[0.    0.57735027    0.57735027    0.    0.57735027    0.    0.    0.]
```

4.5 The Generalized N-qubit Werner state

The two-qubit Werner states are special states in the Hilbert space that are entangled but still admit the local hidden-variable model [58], and they are local unitary invariant states. The density matrix of the two-qubit Werner state is given below,

$$\rho' = p|b_3\rangle\langle b_3| + \frac{1-p}{4}\mathbb{I}_2, \tag{4.43}$$

where $|b_3\rangle = (|00\rangle + |11\rangle)/\sqrt{2}$, and $\mathbb{I}_2$ is an identity matrix of dimension $2^2 \times 2^2$. The generalized N-qubit ($N \geq 3$) Werner [49, 59] state can be written in terms of the N-qubit GHZ state and a N-qubit maximally mixed state, which are mixed suitably with probabilities p and $1 - p$, respectively,

$$\rho(p) = p|GHZ\rangle\langle GHZ| + \frac{1-p}{2^N}\mathbb{I}_N, \tag{4.44}$$

where $\mathbb{I}_N$ is an identity matrix of dimension $2^N \times 2^N$.

```
nWerner(p,tot_spins=2)
```

Parameters

in/out	Argument	Description
[in]	p	It is the mixing probability
[in]	tot_spins	tot_spins stores the total number of qubits
[out]	rho	It stores the density matrix of the Werner state

Implementation

```
def nWerner(self,p,tot_spins=2):
    """

    Construct N-qubit Werner state
    Input:
        tot_spins: it is the total number of spins, it should be equal to
                   or greater than 2.
        p: it is the mixing probability
    Output:
        rho: N-qubit Werner state.
    """
    # Checking total number of spins N >= 2
    assert tot_spins >= 2, "Total number of spins are less than 2"
    qobj=QM()
    # If N = 2 then |GHZ> = |b_3>
    if tot_spins == 2:
        state=self.bell3()
    else:
        state=self.nGHZ(tot_spins=tot_spins)
    den=qobj.outer_product_rvec(state,state)
    # Constructing the identity matrix I_N
    identity=np.identity(2**tot_spins, dtype = 'float64')
    identity=identity*(1/(2**tot_spins))
    # Finally constructing the N-qubit Werner state rho(p)
    rho=(p*den)+((1-p)*identity)
    return rho
```

Example

```
state=gt_obj.nWerner(p=0.5,tot_spins=2)
print(state)
```

The above code prints the two qubit generalized Werner state as given in Eq. [4.43].

```
[[0.375 0.    0.    0.25 ]
 [0.    0.125 0.    0.   ]
 [0.    0.    0.125 0.   ]
 [0.25 0.    0.    0.375]]
```

4.6 Shannon entropy

Shannon entropy is the metric that is used to measure the information entropy of a stream of data bits. The mathematical definition of Shannon entropy is as follows, it quantifies how much information is required on an average to identify random samples from the distribution. Given a discrete random variable X, which has outcomes $\{x_i; i = 1, \cdots N\}$ occurring with corresponding probabilities $\{p_i(x); i = 1, \cdots N\}$, respectively, then the Shannon entropy [60] for the above discrete set of probabilities is given as,

$$H(p) = -\sum_{i=1}^{N} p_i(x) \log_2 p_i(x). \tag{4.45}$$

Shannon entropy is one of the most important information theoretic measures widely used. Note that when $N = 2$, the corresponding entropy is called the binary entropy which is,

$$H_b(p) = -p \log_2 p - (1 - p) \log_2(1 - p). \tag{4.46}$$

Shannon entropy is useful while computing quantities like mutual information, and conditional entropies [61].

`shannon_entropy(pvec)`

Parameters

in/out	Argument	Description
[in]	pvec	It is a real array which contains probabilities
[out]	se	se is the value of Shannon entropy

Implementation

```python
def shannon_entropy(self,pvec):
    """
    Calculates the Shannon entropy
    Input:
        pvec: column vector which contains probabilities
    Output:
        se: it returns the Shannon entropy value
    """
    size=pvec.shape[0]
    se=0.0 # It stores Shannon entropy value H(p)
    for i in range(0,size):
        # Checking probability p is bounded between 0 <= p <= 1
        assert pvec[i]<=1 and pvec[i] >=0, "probability values are
incorrect"
        se=se-(pvec[i]*math.log2(pvec[i]))
    return se
```

Example

In this example, we are finding the Shannon entropy of the probability vector $p = (0.3, 0.2, 0.5)$.

```
pvec=np.array([0.3, 0.2, 0.5])
se=gt_obj.shannon_entropy(pvec)
print(se)
```

The above code prints the required Shannon entropy.

```
1.4854752972273344
```

4.7   Linear entropy

In quantum mechanics, linear entropy is the measure of the impurity or mixedness of state, and it is defined as

$$S_L = 1 - Tr(\rho^2). \tag{4.47}$$

For any state ρ, we know that for pure state $Tr(\rho^2) = 1$, and for a mixed state $Tr(\rho^2) < 1$. The linear entropy in this way is a measure of how much a given state deviates from being a pure state.

4.7.1   For a real or complex density matrix

```
linear_entropy(rho)
```

Parameters

in/out	Argument	Description
[in]	rho	It is a real or complex density matrix
[out]	le	le is the linear entropy of rho

Implementation

```
def linear_entropy(self, rho):
    """
    Calculates the Linear entropy
    Input:
        rho: it is the density matrix
    Output:
        le: linear entropy value
    """
    tr=np.trace(rho)
    # Checking Tr(ρ) = 1
```

```
        assert np.allclose(abs(tr),1), "density matrix is not correct"
        # Calculating Tr(ρ²)
        tr2=np.trace(np.matmul(rho,rho))
        # Finally calculating, S_L = 1 − Tr(ρ²)
        le=1.0-abs(tr2)
        return le
```

Example

In this example, we are finding the linear entropy of a two qubit generalized Werner state $\rho(0)$.

```
state=gt_obj.nWerner(0.0)
le=gt_obj.linear_entropy(state)
print(le)
```

The above code prints the required linear entropy.

```
0.75
```

In the following example, we are finding the linear entropy of a complex density matrix defined below.

$$\rho' = \begin{pmatrix} 0.4 & -0.2 - i0.1 \\ -0.2 + i0.1 & 0.6 \end{pmatrix} \tag{4.48}$$

```
state=np.array([[complex(0.4,0),complex(-0.2,-0.1)],\
                [complex(-0.2,0.1),complex(0.6,0.0)]])
le=gt_obj.linear_entropy(state)
print(le)
```

The above code prints the required linear entropy.

```
0.38
```

There is a term called participation ratio which is also related to purity of state and it is defined as,

$$PR = \frac{1}{Tr(\rho^2)}. \tag{4.49}$$

PR is interpreted as the effective number of pure states entering to form the mixture. We have given an example of finding the participation ratio for $\mathbb{I}/2$.

```
import numpy as np

# Constructing the matrix, I/2
identity_matrix = np.zeros([2,2],dtype='float64')
identity_matrix[0,0]=0.5
identity_matrix[1,1]=0.5

# Calculating the value PR = 1/Tr(ρ²)
PR = 1/np.trace(np.matmul(identity_matrix,identity_matrix))

print(f"The participation ratio is = {PR}")
```

The preceding code generates the following output.

```
The participation ratio is = 2.0
```

4.8   Relative entropy

The closeness between two probability distributions with respect to the quantum states ρ and σ is termed as the relative entropy [24, 62] and it can be defined as

$$S(\rho \parallel \sigma) = Tr(\rho \log \rho) - Tr(\rho \log \sigma). \tag{4.50}$$

Note that there is no upper limit on the value of relative entropy and $S(\rho \parallel \sigma) \geq 0$, the equality holds when $\rho = \sigma$. This definition of relative entropy is with respect to base e, however it can be changed to any base by dividing it with the appropriate conversion factor. Relative entropy is used in the calculation of quantum mutual information, which is an important quantity in information theory.

4.8.1   Between two density matrices

```
relative_entropy(rho,sigma)
```

Parameters

in/out	Argument	Description
[in]	rho	It is a real or complex density matrix
[in]	sigma	It is a real or complex density matrix
[out]	rtent	rtent is the relative entropy between states rho and sigma density matrices

Implementation

```
def relative_entropy(self,rho,sigma):
    """
    Calculates relative entropy
    Input:
        rho: input density matrix
        sigma: input density matrix
    Output:
        rtent: the value of relative entropy
    """
    laobj=LA()
    typerho=str(rho.dtype)
    typesig=str(sigma.dtype)
```

```
if re.findall('^float|^int',typerho):
    # Calculating log(ρ), where ρ = ρ^T
    logrho=laobj.function_smatrix(rho, mode="log",\
                                    log_base=math.exp(1))
elif re.findall("^complex",typerho):
    # Calculating log(ρ), where ρ = ρ^†
    logrho=laobj.function_hmatrix(rho, mode="log",\
                                    log_base=math.exp(1))
if re.findall('^float|^int',typesig):
    # Calculating log(σ), where σ = σ^T
    logsig=laobj.function_smatrix(sigma, mode="log",\
                                    log_base=math.exp(1))
elif re.findall("^complex",typesig):
    # Calculating log(σ), where σ = σ^†
    logsig=laobj.function_hmatrix(sigma, mode="log",\
                                    log_base=math.exp(1))
rtent=np.trace(np.matmul(rho,logrho))-np.trace(np.matmul(rho,logsig))
rtent=abs(rtent)
return rtent
```

Example

In this example, we are finding the relative entropy between the two qubit generalized Werner states $\rho(0.3)$ and $\rho(0.5)$.

```
state1=gt_obj.nWerner(0.3)
state2=gt_obj.nWerner(0.5)
rtent=gt_obj.relative_entropy(state1,state2)
print(rtent)
```

The above code prints the required relative entropy.

```
0.04629042251780047
```

In this example, we are finding the relative entropy between two complex density matrices defined below,

$$\rho_1 = \begin{pmatrix} 0.4 & -0.2 - i0.1 \\ -0.2 + i0.1 & 0.6 \end{pmatrix}, \qquad \rho_2 = \begin{pmatrix} 0.4 & -0.1 - i0.4 \\ -0.1 + i0.4 & 0.6 \end{pmatrix} \qquad (4.51)$$

```
state1=np.array([[complex(0.4,0.0),complex(-0.2,-0.1)],\
                [complex(-0.2,0.1),complex(0.6)]])
state2=np.array([[complex(0.4,0.0),complex(-0.1,-0.4)],\
                [complex(-0.1,0.4),complex(0.6,0.0)]])
rtent=gt_obj.relative_entropy(state1,state2)
print(rtent)
```

The above code prints the required relative entropy.

```
0.3490466018593652
```

4.9 Trace distance

The trace distance [24, 63] is a metric on the space of density matrices and gives a measure of the distinguishability between two states. Trace distance between two density matrices ρ and σ is defined as follows,

$$T(\rho,\sigma) = \frac{1}{2} \parallel \rho - \sigma \parallel. \tag{4.52}$$

From the above, we see that the trace distance is directly related to the trace norm of the difference between the states.

4.9.1 Between two real or complex density matrices

`trace_distance(rho,sigma)`

Parameters

in/out	Argument	Description
[in]	rho	It is a real or complex density matrix of dimension (0:n−1,0:n−1)
[in]	sigma	It is a real or complex density matrix of dimension (0:n−1,0:n−1)
[out]	trd	trd is the trace distance between states rho and sigma

Implementation

```
def trace_distance(self,rho,sigma):
    """
    Calculates trace distance between two density matrices
    Input:
        rho: input density matrix
        sigma: input density matrix
    Output:
        trd: it stores trace distance
    """
    # Calculating ρ − σ
    res=rho-sigma
    laobj=LA()
    typeres=str(res.dtype)
    if re.findall('^float|^int',typeres):
        trd=laobj.trace_norm_rmatrix(res)
        trd=trd/2
    elif re.findall("^complex",typeres):
        trd=laobj.trace_norm_cmatrix(res)
        trd=trd/2
    return trd
```

Example

In this example, we are finding the trace distance between the two-qubit generalized Werner state $\rho(p)$.

```
state1=gt_obj.nWerner(0.3)
state2=gt_obj.nWerner(0.5)
trd=gt_obj.trace_distance(state1,state2)
print(trd)
```

The above code prints the required trace distance.

```
0.14999999999999997
```

In this example, we are finding the trace distance between two complex density matrices ρ_1 and ρ_2 as defined in Eq. [4.51].

```
state1=np.array([[complex(0.4,0.0),complex(-0.2,-0.1)],\
                [complex(-0.2,0.1),complex(0.6)]])
state2=np.array([[complex(0.4,0.0),complex(-0.1,-0.4)],\
                [complex(-0.1,0.4),complex(0.6,0.0)]])
trd=gt_obj.trace_distance(state1,state2)
print(trd)
```

The above code prints the required trace distance.

```
0.31622776601683800
```

4.10 Fidelity

Fidelity [64, 65] is a measure of how close two quantum states are. It has a variety of applications such as in the field of quantum teleportation [66], quantum error correction [67], quantum spin chain system [68], etc. Fidelity between two density matrices ρ and σ can be found using the trace norm as

$$F = \| \sqrt{\rho}\sqrt{\sigma} \|^2 . \tag{4.53}$$

In the special case when ρ and σ are pure states, that is, $\rho = |\psi_\rho\rangle\langle\psi_\rho|$ and $\sigma = |\psi_\sigma\rangle\langle\psi_\sigma|$. Fidelity can be found as

$$F = |\langle\psi_\rho|\psi_\sigma\rangle|^2. \tag{4.54}$$

Fidelity between a pure state $|\psi\rangle$ and a density matrix σ can also be found as

$$F = \langle\psi|\sigma|\psi\rangle. \tag{4.55}$$

Note that fidelity always lies between 0 and 1. In fact the fidelity for pure states is similar to the overlap between the states.

4.10.1 Between two real or complex states

```
fidelity_vec2(self,vecrho,vecsigma)
```

Parameters

in/out	Argument	Description
[in]	vecrho	vecrho is a real or complex state of dimension $(0:2^n-1)$, where n is number of spins or qubits
[in]	vecsigma	vecsigma is a real or complex state of dimension $(0:2^n-1)$, where n is number of spins or qubits
[out]	fidelity	It is the fidelity between states vecrho and vecsigma

Implementation

```python
def fidelity_vec2(self,vecrho,vecsigma):
    """
    Calculates fidelity between two quantum states
    Input:
        vecrho: input pure state vector.
        vecsigma: input pure state vector.
    Output:
        fidelity: it stores the value of fidelity
    """
    typerho=str(vecrho.dtype)
    if re.findall('^complex',typerho):
        # Calculating, ⟨ψ_ρ|ψ_σ⟩
        fidelity=np.matmul(np.matrix.\
                           conjugate(np.matrix.transpose(vecrho)),\
                           vecsigma)
    else:
        # Calculating, ⟨ψ_ρ|ψ_σ⟩
        fidelity=np.matmul(np.matrix.transpose(vecrho), vecsigma)
        # Calculating, |⟨ψ_ρ|ψ_σ⟩|²
    fidelity=abs(fidelity)**2
    return fidelity
```

Example

In this example, we are finding the fidelity between two-qubit states $|\psi_1\rangle$ and $|\psi_2\rangle$.

```python
from QuantumInformation import GatesTools as GT
gt_obj=GT()
state1=np.zeros([4],dtype='float64')
for i in range(0,state1.shape[0]):
    state1[i]=i+1
state1=qobj.normalization_vec(state1)
state2=np.zeros([4],dtype='float64')
state2[0]=1/np.sqrt(2)
state2[3]=1/np.sqrt(2)
F=gt_obj.fidelity_vec2(state1,state2)
print(F)
```

The above code prints the required fidelity.

```
0.4166666666666665
```

In this example, we are finding the fidelity between two-qubit states $|\psi_3\rangle$ and $|\psi_4\rangle$.

```
from QuantumInformation import GatesTools as GT
gt_obj=GT()
state3=np.zeros([4],dtype=np.complex_)
for i in range(0,state3.shape[0]):
    state3[i]=complex(i,i+1)
state3=qobj.normalization_vec(state3)
state4=np.zeros([4],dtype=np.complex_)
for i in range(0,state4.shape[0]):
    state4[i]=complex(i,i+2)
state4=qobj.normalization_vec(state4)
F=gt_obj.fidelity_vec2(state3,state4)
print(F)
```

The above code prints the required fidelity.

```
0.986631016042781
```

4.10.2   Between two real or complex density matrices

```
fidelity_den2(rho,sigma)
```

Parameters

in/out	Argument	Description
[in]	rho	rho is a real or complex density matrix of dimension $(0:2^n-1,0:2^n-1)$, where n is number of spins or qubits
[in]	sigma	sigma is a real or complex state of dimension $(0:2^n-1,0:2^n-1)$, where n is number of spins or qubits
[out]	fidelity	It is the fidelity between density matrices rho and sigma

Implementation

```
def fidelity_den2(self,rho,sigma):
    """
    Calculates fidelity between two density matrices
    Input:
        rho: input density matrix
        sigma: input density matrix
    Output:
        fidelity: it stores the value of fidelity
    """
```

```
laobj=LA()
typerho=str(rho.dtype)
typesig=str(sigma.dtype)
flag=0
if re.findall('^float|^int',typerho):
    # Calculating √ρ, if ρ = ρ^T
    rhosq=laobj.power_smatrix(rho,0.5)
elif re.findall("^complex",typerho):
    # Calculating √ρ, if ρ = ρ†
    rhosq=laobj.power_hmatrix(rho,0.5)
    flag=1
if re.findall('^float|^int',typesig):
    # Calculating √σ, if σ = σ^T
    sigsq=laobj.power_smatrix(sigma,0.5)
elif re.findall("^complex",typesig):
    # Calculating √σ, if σ = σ†
    sigsq=laobj.power_hmatrix(sigma,0.5)
    flag=1
if flag==0:
    # If both matrices are real, calculate ||√ρ√σ||^2
    fidelity=laobj.trace_norm_rmatrix(np.matmul(rhosq,sigsq))
    fidelity=fidelity**2
else:
    # If any of the matrix is complex, calculate ||√ρ√σ||^2
    fidelity=laobj.trace_norm_cmatrix(np.matmul(rhosq,sigsq))
    fidelity=fidelity**2
return fidelity
```

Example

In this example, we are finding the fidelity between two-qubit generalized Werner states $\rho(0.3)$ and $\rho(0.5)$.

```
state1=gt_obj.nWerner(0.3)
state2=gt_obj.nWerner(0.5)
F=gt_obj.fidelity_den2(state1,state2)
print(F)
```

The above code prints the required fidelity.

```
0.9772673859335358
```

In this example, we are finding the fidelity between the complex states ρ_1 and ρ_2.

```
state1=np.array([[complex(0.4,0.0),complex(-0.2,-0.1)],\
                 [complex(-0.2,0.1),complex(0.6)]])

state2=np.array([[complex(0.4,0.0),complex(-0.1,-0.4)],\
                 [complex(-0.1,0.4),complex(0.6,0.0)]])
F=gt_obj.fidelity_den2(state1,state2)
print(F)
```

The above code prints the required fidelity.

0.8706512518934156

4.10.3 Between a state and a density matrix

`fidelity_vecden(vec,sigma)`

Parameters

in/out	Argument	Description
[in]	vec	vec is a real or complex state of dimension $(0:2^n-1)$, where n is number of spins or qubits
[in]	sigma	sigma is a real or complex state of dimension $(0:2^n-1,0:2^n-1)$, where n is number of spins or qubits
[out]	fidelity	It is the fidelity between state vec and density matrix sigma

Implementation

```python
def fidelity_vecden(self,vec,sigma):
    """
    Calculates fidelity between a quantum state and a density matrix
    Input:
        vec: input pure state vector.
        sigma: input density matrix
    Output:
        fidelity: it stores the value of fidelity
    """
    typevec=str(vec.dtype)
    if re.findall('^complex',typevec):
        # Complex case: ⟨ψ|σ|ψ⟩
        fidelity=np.matmul(np.matrix.\
                        conjugate(np.matrix.transpose(vec)),\
                        np.matmul(sigma,vec))
    else:
        # Real case: ⟨ψ|σ|ψ⟩
        fidelity=np.matmul(np.matrix.transpose(vec),\
                        np.matmul(sigma,vec))
    fidelity=abs(fidelity)
    return fidelity
```

Example

In this example, we are finding the fidelity between two-qubit state $|\psi_2\rangle$ and two-qubit generalized Werner state $\rho(0.3)$.

```
from QuantumInformation import GatesTools as GT
gt_obj=GT()
state1=np.zeros([4],dtype='float64')
for i in range(0,4):
    state1[i]=i+1
state1=qobj.normalization_vec(state1)
state2=gt_obj.nWerner(0.3)
F=gt_obj.fidelity_vecden(state1,state2)
print(F)
```

The above code prints the required fidelity.

```
0.29999999999999993
```

In this example, we are finding the fidelity between the one-qubit complex state $|\psi_3\rangle$ and complex density matrix ρ_1.

```
from QuantumInformation import GatesTools as GT
gt_obj=GT()
state1=np.zeros([2],dtype=np.complex_)
for i in range(0,2):
    state1[i]=complex(i,i+1)
state1=qobj.normalization_vec(state1)
state2=np.array([[complex(0.4,0.0),complex(-0.2,-0.1)],\
                [complex(-0.2,0.1),complex(0.6,0.0)]])
F=gt_obj.fidelity_vecden(state1,state2)
print(F)
```

The above code prints the required fidelity.

```
0.4
```

4.11   Super fidelity

Super fidelity [69] is an upper bound on the fidelity of two quantum states which is defined by

$$G(\rho,\sigma) = Tr(\rho\sigma) + \sqrt{(1-Tr(\rho^2)(1-Tr(\sigma^2)},\qquad(4.56)$$

for two-qubit states the super fidelity is the same as fidelity. The super fidelity is bounded between 0 and 1, and it is concave function as shown below,

$$G\left(\rho_1, x\rho_2 + (1-x)\rho_3\right) \geq x\ G\left(\rho_1, \rho_2\right) + (1-x)G\left(\rho_1, \rho_3\right)\qquad(4.57)$$

In its own right, super fidelity can be used as an entanglement measure.

4.11.1   Between two real or complex density matrices

```
super_fidelity(rho,sigma)
```

Parameters

in/out	Argument	Description
[in]	rho	rho is a real or complex density matrix of dimension $(0{:}2^n-1, 0{:}2^n-1)$, where n is number of spins or qubits
[in]	sigma	sigma is a real or complex density matrix of dimension $(0{:}2^n-1, 0{:}2^n-1)$, where n is number of spins or qubits
[out]	sf	It is the super-fidelity between density matrices rho and sigma

Implementation

```python
def super_fidelity(self,rho,sigma):
    """
    Calculates super fidelity between two density matrices
    Input:
        rho: input density matrix.
        sigma: input density matrix.
    output:
        sf: the value of the super fidelity
    """
    # Calculating Tr(ρ²)
    tr_rho2=np.trace(np.matmul(rho,rho))
    # Calculating Tr(σ²)
    tr_sigma2=np.trace(np.matmul(sigma,sigma))
    # Calculating Tr(ρσ)
    tr_rhosigma=np.trace(np.matmul(rho,sigma))
    # Below we calculate, G(ρ,σ)
    sf=tr_rhosigma+np.sqrt((1-tr_rho2)*(1-tr_sigma2))
    sf=abs(sf)
    return sf
```

Example

In this example, we are finding the super fidelity between the two-qubit generalized Werner states $\rho(0.3)$ and $\rho(0.5)$.

```python
state1=gt_obj.nWerner(0.3)
state2=gt_obj.nWerner(0.5)
SF=gt_obj.super_fidelity(state1,state2)
print(SF)
```

The above code prints the required super fidelity.

```
0.9821016865696865
```

In this example, we are finding the super fidelity between two complex density matrices ρ_1 and ρ_2.

```
state1=np.array([[complex(0.4,0.0),complex(-0.2,-0.1)],\
                [complex(-0.2,0.1),complex(0.6)]])

state2=np.array([[complex(0.4,0.0),complex(-0.1,-0.4)],\
                [complex(-0.1,0.4),complex(0.6,0.0)]])
SF=gt_obj.super_fidelity(state1,state2)
print(SF)
```

The above code prints the required super fidelity.

```
0.870651251893416
```

4.12   Bures distance

Bures distance [70–72] is defined as an infinitesimal distance between density matrix operators defining quantum states. Bures distance between two quantum states ρ and σ can be defined as,

$$D = \sqrt{2(1 - \sqrt{F(\rho,\sigma)})},$$ (4.58)

where F is the fidelity between two quantum states ρ and σ. In its own right Bures distance qualifies to be an entanglement measure.

4.12.1   Between two real or complex states

```
bures_distance_vec(rho,sigma)
```

Parameters

in/out	Argument	Description
[in]	rho	rho is a real or complex state of dimension $(0:2^n-1)$, where n is number of spins or qubits
[in]	sigma	sigma is a real or complex state of dimension $(0:2^n-1)$, where n is number of spins or qubits
[out]	bd	It is the Bures distance between the states rho and sigma

Implementation

```
def bures_distance_vec(self,rho,sigma):
    """
    Calculates Bures distance between two quantum state
    Input:
        rho: input state vector
        sigma: input state vector
```

```
    Output:
        bd: the value of the Bures distance
    """
    # Calculating F(ρ,σ)
    fid=self.fidelity_vec2(rho,sigma)
    # Finally we calculate Bures distance D
    bd=np.sqrt(2*(1-np.sqrt(fid)))

    return bd
```

Example

In this example, we are finding the Bures distance between the two-qubit states $|\psi_1\rangle$ and $|\psi_2\rangle$.

```
from QuantumInformation import GatesTools as GT
gt_obj=GT()
state1=np.zeros([4],dtype='float64')
state2=np.zeros([4],dtype='float64')
for i in range(0,4):
    state1[i]=i+1
state1=qobj.normalization_vec(state1)
state2[0]=1/np.sqrt(2)
state2[3]=1/np.sqrt(2)
Bd=gt_obj.bures_distance_vec(state1,state2)
print(Bd)
```

The preceding code prints the required Bures distance.

```
0.8420246737858663
```

In this example, we are finding the Bures distance between the two-qubit complex states $|\psi_3\rangle$ and $|\psi_4\rangle$.

```
state1=np.zeros([4],dtype=np.complex_)
state2=np.zeros([4],dtype=np.complex_)
for i in range(0,4):
    state1[i]=complex(i,i+1)
    state2[i]=complex(i,i+2)
state1=qobj.normalization_vec(state1)
state2=qobj.normalization_vec(state2)
Bd=gt_obj.bures_distance_vec(state1,state2)
print(Bd)
```

The preceding code prints the required Bures distance.

```
0.11581868410941999
```

4.12.2 Between two real or complex density matrices

```
bures_distance_den(rho,sigma)
```

Parameters

in/out	Argument	Description
[in]	rho	rho is a real or complex density matrix of dimension $(0:2^n-1,0:2^n-1)$, where n is number of spins or qubits
[in]	sigma	sigma is a real or complex density matrix of dimension $(0:2^n-1,0:2^n-1)$, where n is number of spins or qubits
[out]	bd	It is the Bures distance between the density matrices rho and sigma

Implementation

```python
def bures_distance_den(self,rho,sigma):
    """
    Calculates Bures distance between two density matrix
    Input:
        rho: input density matrix
        sigma: input density matrix
    Output:
        bd: the value of the Bures distance
    """
    # Calculating F(ρ,σ)
    fid=self.fidelity_den2(rho,sigma)
    # Finally we calculate Bures distance D
    bd=np.sqrt(2*(1-np.sqrt(fid)))

    return bd
```

Example

In this example, we are finding the Bures distance between the two qubit generalized Werner states $\rho(0.3)$ and $\rho(0.5)$.

```python
state1=gt_obj.nWerner(0.3)
state2=gt_obj.nWerner(0.5)
Bd=gt_obj.bures_distance_den(state1,state2)
print(Bd)
```

The preceding code prints the required Bures distance.

```
0.15120613959059134
```

In this example, we are finding the Bures distance between complex states ρ_1 and ρ_2.

```python
state1=np.array([[complex(0.4,0.0),complex(-0.2,-0.1)],\
                [complex(-0.2,0.1),complex(0.6)]])

state2=np.array([[complex(0.4,0.0),complex(-0.1,-0.4)],\
                [complex(-0.1,0.4),complex(0.6,0.0)]])
```

```
Bd=gt_obj.bures_distance_den(state1,state2)
print(Bd)
```

The preceding code prints the required Bures distance.

```
0.3658225043378863
```

4.13 Expectation value of an observable

In quantum mechanics, the expectation value [24, 73, 74] of any observable A with respect to pure state $|\psi\rangle$ is defined as

$$\langle A \rangle = \langle \psi | A | \psi \rangle. \tag{4.59}$$

The above equation also can be written in terms of the density matrix representation ρ as,

$$\langle A \rangle = Tr(\rho A). \tag{4.60}$$

When many measurements are done on N copies of the quantum system, the expectation value is the average value obtained when the number of such systems goes to infinity, that is, $N \to \infty$. The expectation value of any observable is a real number.

4.13.1 For a real or complex state

```
expectation_vec(vec,obs)
```

Parameters

in/out	Argument	Description
[in]	vec	It is a real or complex array of dimension $(0{:}n{-}1)$
[in]	obs	It is the observable matrix of dimension $(0{:}n{-}1,0{:}n{-}1)$
[out]	expc	expc is the expectation value of obs corresponding to the state vec

Implementation

```
def expectation_vec(self,vec,obs):
    """
    Expectation values of observable for a quantum state
    Input:
        vec: input state vector
        obs: observable operator
    Output:
        expc: the expectation value of the measurement operator
    """
```

```
    typevec=str(vec.dtype)
    if re.findall('^complex',typevec):
        # Calculating ⟨ψ|A|ψ⟩, if |ψ⟩ is complex
        expc=np.matmul(np.matmul(np.matrix.\
                        conjugate(np.matrix.transpose(vec)),\
                        obs),vec)
    else:
        # Calculating ⟨ψ|A|ψ⟩, if |ψ⟩ is real
        expc=np.matmul(np.matmul(np.matrix.transpose(vec),obs),vec)
    return expc
```

Example

In this example, we are finding the expectation value of σ^z with respect to $|1\rangle$.

```
sz=gt_obj.sz()
state=np.array([0,1])
expec=gt_obj.expectation_vec(state,sz)
print(expec)
```

The above code prints the required expectation value.

```
-1
```

Another example where we are finding the expectation value of σ^y with respect to $\frac{|0\rangle - i|1\rangle}{2}$.

```
sy=gt_obj.sy()
state=np.array([1/np.sqrt(2),complex(0,-1/np.sqrt(2))])
expec=gt_obj.expectation_vec(state,sy)
print(expec)
```

The above code prints the required expectation value.

```
-0.9999999999999998
```

4.13.2   For a real or a complex density matrix

```
expectation_den(rho,obs)
```

Parameters

in/out	Argument	Description
[in]	rho	It is a real or complex array of dimension $(0{:}n{-}1,0{:}n{-}1)$
[in]	obs	It is the observable matrix of dimension $(0{:}n{-}1,0{:}n{-}1)$
[out]	expc	expc is the expectation value of obs corresponding to the state rho

Implementation

```
def expectation_den(self,rho,obs):
    """
    Expectation values of observable for a density matrix
    Input:
        rho: input density matrix
        obs: observable operator
    Output:
        expc: the expectation value of the observable operator
    """
    return np.trace(np.matmul(rho,obs))
```

Example

In this example, we are finding the expectation value of σ^z with respect to $|0\rangle\langle 0|$.

```
sz=gt_obj.sz()
rho=np.array([[1,0],[0,0]])
expc=gt_obj.expectation_den(rho,sz)
print(expc)
```

The preceding code prints the required expectation value.

```
1
```

Another example where we are finding the expectation value of σ^y with respect to $\left(\frac{|0\rangle+i|1\rangle}{\sqrt{2}}\right)\left(\frac{\langle 0|-i\langle 1|}{\sqrt{2}}\right)$.

```
sy=gt_obj.sy()
state=np.array([1/np.sqrt(2),complex(0,-1/np.sqrt(2))])
state=qobj.outer_product_cvec(state,state)
expc=gt_obj.expectation_den(state,sy)
print(expc)
```

The preceding code prints the required expectation value.

```
0.9999999999999998
```

4.14 Complete example

In this section, we give an example of constructing some quantum objects using the Python codes. To this end, we construct the quantum state $|\phi_1\rangle$ for $N = 4$ qubits (4 qubit Bell states).

$$|\phi_1\rangle_4 = \frac{1}{2}\left(|01\rangle_{12} + |10\rangle_{12}\right) \otimes \left(|01\rangle_{34} + |10\rangle_{34}\right). \tag{4.61}$$

Thereafter, we construct 4-qubit Werner state, with the mixing probability set to 0.5.

$$\rho_4' = 0.5\left(\frac{|0000\rangle + |1111\rangle}{\sqrt{2}}\right)\left(\frac{\langle 0000| + \langle 1111|}{\sqrt{2}}\right) + \frac{0.5}{16}\mathbb{I}_{16\times 16}, \tag{4.62}$$

where $\mathbb{I}_{16\times16}$ is a 16×16 identity matrix. Finally, we calculate the fidelity between the pure state $|\phi_1\rangle_4$, and density matrix ρ'_4 as shown below.

$$F = {}_4\langle\phi_1|\rho'_4|\phi_1\rangle_4. \tag{4.63}$$

The code below accomplishes the above tasks,

```
from QuantumInformation import GatesTools as GT
gt_obj=GT()

# Constructing the state |φ₁⟩₄
state=gt_obj.bell1(4)

# Constructing the density matrix ρ'₄
den=gt_obj.nWerner(0.5,tot_spins=4)

# Finally calculating fidelity, F = ₄⟨φ₁|ρ'₄|φ₁⟩₄
fidelity=gt_obj.fidelity_vecden(state,den)

print(f"The fidelity is calculated out to be {fidelity}")
```

For the preceding code we obtain the following result.

```
The fidelity is calculated out to be 0.031249999999999986
```

5

Quantification of Quantum Entanglement

Quantum entanglement [75–79] is an exclusive property of quantum states, such a phenomenon is not there in classical mechanics. However, presenting a classical version of entanglement will enable us to understand the characteristic feature of quantum entanglement. Entanglement arises due to a lack of independence between two systems. For example, consider a random variable X, which has two possible outcomes, head (**H**) and tail (**T**). The discrete probability distribution of two-coin flips is given below, where the outcomes of each coin are independent of the other.

Coin 2

	H	T
H	0.25	0.25
T	0.25	0.25

Coin 1

FIGURE 5.1: Discrete probability distribution of two unbiased coin flips.

From the probability distribution shown in Fig. (5.1), it is clear that the outcome of any one of the coins does not influence the outcome of the other coin. If the outcome of the first coin is head, then with a 50% probability that the outcome of the second coin will be head or tail. Now consider a case where we have two special coins, which have discrete probability distributions as shown in Fig. (5.2). The table in Fig. (5.2) shows extreme entanglement.

Coin 2

	H	T
H	0.5	0
T	0	0.5

Coin 1

FIGURE 5.2: Discrete probability distribution of two special coin flips.

If the outcome of the first special coin is head, then with 100% surety, we know the outcome of the second special coin will be head. With this example discussed we now introduce a more formal definition of quantum entanglement.

Entanglement is a non-local correlation which exists between systems that are separated in space. Any measurement in one of the subsystems will affect the state of the

DOI: 10.1201/9781003285489-5

other subsystem, essentially it means that the subsystems are no longer independent [58]. Entanglement plays an important role in not only understanding the basic nature of quantum states but also indispensable in understanding condensed matter systems like spin chains [80–82], and quantum phase transitions [83, 84]. It is also an integral part of quantum computation [85–87], etc.

The nature of entanglement in pure and mixed states are very different from each other and, as we go along, we shall see the ways to detect and quantify entanglement. As far as pure states are concerned, if we have a composite system AB which is in a state $|\psi\rangle_{AB}$ made of two subsystems A and B, if the state of the composite system can be written as a tensor product of the individual subsystem states, then we say that $|\psi\rangle_{AB}$ unentangled, if not they are entangled. For example, consider one of the well known two-qubit Bell states which is,

$$|b_1\rangle \;=\; \frac{1}{\sqrt{2}}(|00\rangle_{AB} + |11\rangle_{AB})$$

$$\neq \;\; |\text{some state of system A}\rangle \otimes |\text{some state of system B}\rangle \qquad (5.1)$$

In the above equation, we see that $|b_1\rangle$ is an entangled state. However, in the same token, it is easy to see that the state $|\phi\rangle = \frac{1}{\sqrt{2}}(|00\rangle_{AB} + |01\rangle_{AB}) = \frac{1}{\sqrt{2}}(|0\rangle_A \otimes (|0\rangle_B + |1\rangle_B))$ is an unentangled state. In the state defined as $|b_1\rangle$, we see that it is defined in terms of the eigenvectors of the Pauli Z matrix, however, if we rotate the basis such that the state is written in terms of the eigen vectors of the Pauli X matrix, even then, the non-local correlations or entanglement does not die as the new state preserves these correlations up to an arbitrary global phase. In the language of measurement theory, we see that in the state $|b_1\rangle$, if we measure the spin in the first qubit to be in the state $|0\rangle$ we can immediately, without even measuring the second qubit say with 100 percent probability that it will be in the state $|0\rangle$, however, if we see the state $|\phi\rangle$, measuring the first qubit to be in state $|0\rangle$, will throw the state of the second qubit in a superposition namely $\frac{1}{\sqrt{2}}(|0\rangle_B + |1\rangle_B)$, which is a hallmark of entanglement. However, coming to the arena of density matrices, A density matrix ρ is called separable [88] if it can be written as a convex sum of product states ρ^A and ρ^B, as

$$\rho = \sum_i p_i \rho_i^A \otimes \rho_i^B, \qquad (5.2)$$

This definition, coupled with the fact that there is no unique decomposition [89] for a given density matrix in terms of pure states, leads to a huge market of entanglement measures and associated entanglement detection techniques. Some of them like PPT and reduction are commonly used, and day by day, the research on these increase the number of such measures and detection methods which practically makes them weaker or stronger with respect to each other [90]. For qubit-qubit and qubit-qutrit systems (2×2 and 2×3, respectively), the simple Peres–Horodecki criterion or the famous positive partial transpose PPT criteria [88] provides both a necessary and a sufficient criterion for separability. This cannot be generalized to higher-dimensional quantum states, directly makes detection non-unique.

For studying the entanglement content [91] or detecting the presence of entanglement between two subsystems of a quantum system, we perform the following mathematical operations on the quantum state of the composite quantum system, partial trace and partial transpose. More detailed discussions regarding partial trace and partial transpose are given in Sections 5.1 and 5.2, respectively.

In this chapter, we will be using the same states and density matrices as given in Chapter 4 to demonstrate our recipes and outputs. (In this chapter) Here, the array named "site" stores the index corresponding to the site of the qubits, which has been done via zero-based indexing, meaning 0 corresponds to the first qubit and so on. Primarily we will be discussing two classes, **PartialTr** and **Entanglement**. Both of these classes are written inside the Python module **chap5_partial_quantumentanglement.py**. The class **PartialTr** contains all the methods pertaining to partial trace and partial transpose. The **Entanglement** class also inherits all methods from the **PartialTr** class. The procedure for importing the classes and creating their objects, you have to write the following code,

```
# Importing both of the classes
from QuantumInformation import PartialTr as PT
from QuantumInformation import Entanglement as ENT

# Creating the objects
pobj=PT()
entobj=ENT()
```

5.1 Partial trace

Partial trace [24, 92] is a mathematical operation that is not only used to study the state of the subsystems of a composite quantum system but also used in the computation of a vast number of entanglement measures. Consider H represents the Hilbert space of a composite quantum system that comprises two subsystems. The Hilbert spaces of the subsystems A and B are denoted by H_A and H_B, respectively. The state of the composite quantum system is known, and it is represented by ρ_{AB}. On performing partial trace operation on the state ρ_{AB}, we can calculate the reduced density matrices of the subsystems A and B represented by ρ_A and ρ_B, respectively. The mathematical expression to calculate the reduced density matrices ρ_A and ρ_B are shown below.

$$\rho_A = Tr_B(\rho_{AB}). \tag{5.3}$$

Similarly, state of the subsystem B which is ρ_B is given by,

$$\rho_B = Tr_A(\rho_{AB}). \tag{5.4}$$

It can be generalized to multi-qubit systems also in this way, if we have a N-qubit state $\rho_{123\cdots N}$, and if we are interested to know the state of the even qubit subsystem, we can write $\rho_{246\cdots} = Tr_{135\cdots}(\rho_{123\cdots N})$. This implies that partial tracing can be done over a random set of qubits. The partial trace operation is the faithful reflection of the marginal of a multivariate probability distribution in the quantum domain. Partial trace operation is practically used in calculating most of the entanglement measures as we will see further down. In the recipes which we have given, for a N-qubit state, we can calculate the partial trace over any random set of qubits as $\rho_{i_1 i_2 \cdots i_N}$.

The following two methods use the function **recur_comb_add** from the module **RecurNum** and two private methods **___dectobin**, **___bintodec**, which are present within the **PartialTr** class. The private methods are those methods which cannot be accessed outside of the class. The reason to develop private methods is explained as follows, the private method **___dectobin** (**___bintodec**) converts a decimal (binary) number to an equivalent binary (decimal) number. We have already developed methods that convert decimal to binary and vice versa, but the format of the input and output of these private methods are different. The algorithm for the partial tracing in the following two methods has been adopted from the ref. [92], and it is the fastest algorithm to calculate partial trace reported till now to the best of our knowledge. The ref. [92] uses the concept of power set to construct the reduced density matrix, and to this end, we have used the recursive function **recur_comb_add** in order to generate the power set.

5.1.1   For a real or complex state

---

partial_trace_vec(state,sub_tr)

Parameters

In/Out	Argument	Description
[in]	state	It is a real or complex array of dimension $(0:2^N-1)$, which is the input state
[in]	sub_tr	It is a list of numbers designating the particular subsystems not to be traced out. The list can store any natural number between 1 and N
[out]	red_den	It is a real or complex array of dimension $(0:2^L-1, 0:2^L-1)$, where L are the total number of remaining qubits after partial tracing. red_den stores the desired reduced density matrix

Implementation

```python
def partial_trace_vec(self,state,sub_tr):
    """
    Partial trace operation on a quantum state
    Input:
        state: real state vector
        sub_tr: details of the subsystems not to be traced out
    Output:
        red_den: reduced density matrix
    """
    # Storing data type (real or complex) input vector
    typestate=str(state.dtype)
    # Storing the number of spins
    N=int(math.log2(state.shape[0]))
    # Number of subsystems to be traced out
    length=len(sub_tr)
    # Checking whether entered subsystems are valid or not
    assert set(sub_tr).issubset(set(np.arange(1,N+1))),\
    "Invalid subsystems to be traced out"
    # If the input state is complex then following condition is true
    if re.findall("^complex",typestate):
        # Constructing the reduced density matrix ρ'
        red_den=np.zeros([2**(length),2**(length)],dtype=np.complex_)
        vec=np.zeros([(N-length),1])
```

```
    im=0
    # Binary place value of the traced out subsystems
    for ii in range(1,N+1):
        if ii not in sub_tr:
            # Creating the set S
            vec[im]=2**(N-ii)
            im=im+1
    mylist=[]
    icount=0
    sum2=0
    # Creating the power set P(S)
    RecurNum.recur_comb_add(mylist,vec,icount,sum2)
    irow=np.zeros([N,1])
    icol=np.zeros([N,1])
    mylist=np.array(mylist)
    len_mylist=len(mylist)
    # Row index of ρ'
    for i1 in range(0,2**length):
        col1=self.__dectobin(i1,length)
        # Column index of ρ'
        for i2 in range(0,2**length):
            col2=self.__dectobin(i2,length)
            i3=0
            for k in range(0,N):
                if k+1 not in sub_tr:
                    irow[k]=0
                else:
                    irow[k]=col1[i3]
                    i3=i3+1
            ic=0
            for k2 in range(0,N):
                if k2+1 not in sub_tr:
                    icol[k2]=0
                else:
                    icol[k2]=col2[ic]
                    ic=ic+1
            icc=self.__bintodec(irow)
            jcc=self.__bintodec(icol)
            red_den[i1,i2]=red_den[i1,i2]+(state[icc]*\
                    np.conjugate(state[jcc]))
            for jj in range(0,len_mylist):
                icc2=icc+mylist[jj]
                jcc2=jcc+mylist[jj]
                red_den[i1,i2]=red_den[i1,i2]+(state[icc2]*\
                        np.conjugate(state[jcc2]))
# If the input state is real then following condition is true
else:
    # Constructing the reduced density matrix ρ'
    red_den=np.zeros([2**(length),2**(length)],dtype='float64')
    vec=np.zeros([(N-length),1])
    im=0
```

```
# Binary place value of the traced out subsystems
for ii in range(1,N+1):
    if ii not in sub_tr:
        # Creating the set S
        vec[im]=2**(N-ii)
        im=im+1
mylist=[]
icount=0
sum2=0
# Creating the power set P(S)
RecurNum.recur_comb_add(mylist,vec,icount,sum2)
irow=np.zeros([N,1])
icol=np.zeros([N,1])
mylist=np.array(mylist)
len_mylist=len(mylist)
# Row index of ρ'
for i1 in range(0,2**length):
    col1=self.__dectobin(i1,length)
    # Column index of ρ'
    for i2 in range(0,2**length):
        col2=self.__dectobin(i2,length)
        i3=0
        for k in range(0,N):
            if k+1 not in sub_tr:
                irow[k]=0
            else:
                irow[k]=col1[i3]
                i3=i3+1
        ic=0
        for k2 in range(0,N):
            if k2+1 not in sub_tr:
                icol[k2]=0
            else:
                icol[k2]=col2[ic]
                ic=ic+1
        icc=self.__bintodec(irow)
        jcc=self.__bintodec(icol)
        red_den[i1,i2]=red_den[i1,i2]+(state[icc]*state[jcc])
        for jj in range(0,len_mylist):
            icc2=icc+mylist[jj]
            jcc2=jcc+mylist[jj]
            red_den[i1,i2]=red_den[i1,i2]+(state[icc2]*state[jcc2])
return(red_den)
```

Example

In this example, we are performing the partial trace on the real four-qubit state $|\psi_2\rangle$ and we will get the reduced density matrix ρ_{24} as an output by tracing over qubits 1 and 3.

```
from QuantumInformation import QuantumMechanics as QM
qobj=QM()
# state1 will store |ψ₂⟩
state1=np.zeros([16],dtype='float64')
for i in range(0,16):
    state1[i]=i+1
# Finally constructing the state |ψ₂⟩
state1=qobj.normalization_vec(state1)
# The subsystems to be retained
sub_tr=[2,4]
# Finally obtaining ρ₂₄
red_den=pobj.partial_trace_vec(state1,sub_tr)
print(red_den)
```

The preceding code prints the required reduced density matrix ρ_{24} as shown below.

0.141711235	0.157754004	0.205882356	0.221925139
0.157754004	0.176470593	0.232620314	0.251336902
0.205882356	0.232620314	0.312834233	0.339572191
0.221925139	0.251336902	0.339572191	0.368983954

Similarly, we perform partial trace on the complex four-qubit state $|\psi_3\rangle$ and we will obtain the reduced density matrix ρ_4 as an output by tracing over qubits 1, 2 and 3.

```
from QuantumInformation import QuantumMechanics as QM
qobj=QM()
# state2 will store |ψ₃⟩
state2=np.zeros([16],dtype=np.complex_)
for i in range(0,16):
    state2[i]=complex(i,i+1)
# Finally constructing the state |ψ₃⟩
state2=qobj.normalization_vec(state2)
# The subsystems to be retained
sub_tr=[4]
# Finally obtaining ρ₄
red_den=pobj.partial_trace_vec(state2,sub_tr)
print(red_den)
```

The preceding code prints the required reduced density matrix ρ_4 as shown below.

```
[[0.45321637+0.j         0.49707602+0.00292398j]
 [0.49707602-0.00292398j 0.54678363+0.j        ]]
```

5.1.2   For a real or complex density matrix

```
partial_trace_den(state,sub_tr)
```

Parameters

In/Out	Argument	Description
[in]	state	It is a real or complex array of dimension $(0:2^N-1,0:2^N-1)$, which is the input density matrix
[in]	sub_tr	It is list of numbers designating the particular subsystems not to be traced out. The list can store any natural number between 1 and N
[out]	red_den	It is a real or complex array of dimension $(0:2^L-1,0:2^L-1)$, where L are the total number of remaining qubits after partial tracing. red_den stores the desired reduced density matrix

Implementation

```python
def partial_trace_den(self,state,sub_tr):
    """
    Partial trace operation on a density matrix
    Input:
        state: input real density matrix
        sub_tr: details of the subsystem not to be traced out
    Output:
        red_den: reduced density matrix
    """
    # Data type (real or complex) of the input density matrix
    typestate=str(state.dtype)
    N=int(math.log2(state.shape[0]))
    length=len(sub_tr)
    # count=length, and count0= N-length
    # Checking whether entered subsystems are valid or not
    assert set(sub_tr).issubset(set(np.arange(1,N+1))),\
    "Invalid subsystems to be traced out"
    # If the density matrix is complex then following condition is true
    if re.findall("^complex",typestate):
        # Constructing the reduced density matrix ρ'
        red_den=np.zeros([2**(length),2**(length)],dtype=np.complex_)
        vec=np.zeros([(N-length),1])
        im=0
        for ii in range(1,N+1):
            if ii not in sub_tr:
                # Constructing the set S
                vec[im]=2**(N-ii)
                im=im+1
        mylist=[]
        icount=0
        sum2=0
        # Constructing the power set P(S)
        RecurNum.recur_comb_add(mylist,vec,icount,sum2)
```

```
            irow=np.zeros([N,1])
            icol=np.zeros([N,1])
            mylist=np.array(mylist)
            len_mylist=len(mylist)
            # Row index of ρ'
            for i1 in range(0,2**length):
                col1=self.__dectobin(i1,length)
                # Column index of ρ'
                for i2 in range(0,2**length):
                    col2=self.__dectobin(i2,length)
                    i3=0
                    for k in range(0,N):
                        if k+1 not in sub_tr:
                            irow[k]=0
                        else:
                            irow[k]=col1[i3]
                            i3=i3+1
                    ic=0
                    for k2 in range(0,N):
                        if k2+1 not in sub_tr:
                            icol[k2]=0
                        else:
                            icol[k2]=col2[ic]
                            ic=ic+1
                    icc=self.__bintodec(irow)
                    jcc=self.__bintodec(icol)
                    red_den[i1,i2]=red_den[i1,i2]+(state[icc,jcc])
                    for jj in range(0,len_mylist):
                        icc2=icc+mylist[jj]
                        jcc2=jcc+mylist[jj]
                        red_den[i1,i2]=red_den[i1,i2]+(state[icc2,jcc2])
# If the density matrix is real then following condition is true
else:
    # Constructing the reduced density matrix ρ'
    red_den=np.zeros([2**(length),2**(length)],dtype='float64')
    vec=np.zeros([(N-length),1])
    im=0
    for ii in range(1,N+1):
        if ii not in sub_tr:
            # Constructing the set S
            vec[im]=2**(N-ii)
            im=im+1
    mylist=[]
    icount=0
    sum2=0
    # Constructing the power set P(S)
    RecurNum.recur_comb_add(mylist,vec,icount,sum2)
    irow=np.zeros([N,1])
    icol=np.zeros([N,1])
    mylist=np.array(mylist)
    len_mylist=len(mylist)
```

```
            # The row index of the ρ'
        for i1 in range(0,2**length):
            col1=self.__dectobin(i1,length)
            # The column index of the ρ'
            for i2 in range(0,2**length):
                col2=self.__dectobin(i2,length)
                i3=0
                for k in range(0,N):
                    if k+1 not in sub_tr:
                        irow[k]=0
                    else:
                        irow[k]=col1[i3]
                        i3=i3+1
                ic=0
                for k2 in range(0,N):
                    if k2+1 not in sub_tr:
                        icol[k2]=0
                    else:
                        icol[k2]=col2[ic]
                        ic=ic+1
                icc=self.__bintodec(irow)
                jcc=self.__bintodec(icol)
                red_den[i1,i2]=red_den[i1,i2]+(state[icc,jcc])
                for jj in range(0,len_mylist):
                    icc2=icc+mylist[jj]
                    jcc2=jcc+mylist[jj]
                    red_den[i1,i2]=red_den[i1,i2]+(state[icc2,jcc2])
    return(red_den)
```

Example

In this example, we are performing the partial trace on the real four-qubit state $|\psi_2\rangle\langle\psi_2|$ and we will get the reduced density matrix ρ_{24} as an output by tracing out qubits 1 and 3.

```
from QuantumInformation import QuantumMechanics as QM
qobj=QM()
# state2 will store |ψ₂⟩
state2=np.zeros([2**4],dtype='float64')
for i in range(0,state2.shape[0]):
    state2[i]=i+1
# Finally constructing the state |ψ₂⟩
state2=qobj.normalization_vec(state2)
# Constructing the density matrix, |ψ₂⟩⟨ψ₂|
den=qobj.outer_product_rvec(state2,state2)
sub_tr=[2,4]
# Finally calculated the reduced density matrix, ρ₂₄
red_den=pobj.partial_trace_den(den,sub_tr)
print(red_den)
```

The preceding code prints the required reduced density matrix ρ_{24} as shown below.

```
[[0.14171123 0.15775401 0.20588235 0.22192513]
 [0.15775401 0.17647059 0.23262032 0.2513369 ]
 [0.20588235 0.23262032 0.31283422 0.33957219]
 [0.22192513 0.2513369  0.33957219 0.36898396]]
```

Similarly, we perform the partial trace on a four-qubit state $|\psi_3\rangle\langle\psi_3|$ and we will get the reduced density matrix ρ_4 as an output by tracing over qubits 1, 2 and 3.

```
from QuantumInformation import QuantumMechanics as QM
qobj=QM()
# state2 will store |ψ3⟩
state2=np.zeros([2**4],dtype=np.complex_)
for i in range(0,state2.shape[0]):
    state2[i]=complex(i,i+1)
# Finally constructing the state |ψ3⟩
state2=qobj.normalization_vec(state2)
# Constructing the density matrix, |ψ3⟩⟨ψ3|
den=qobj.outer_product_cvec(state2,state2)
sub_tr=[4]
# Finally calculated the reduced density matrix, ρ4
red_den=pobj.partial_trace_den(den,sub_tr)
print(red_den)
```

The preceding code prints the required reduced density matrix ρ_4 as shown below.

```
[[0.45321637+0.j         0.49707602+0.00292398j]
 [0.49707602-0.00292398j 0.54678363+0.j         ]]
```

5.2   Partial transpose

Partial transpose [88, 93] is a very important operation in checking the separability of a given density matrix. There are many separability tests that use partial transpose in order to check whether a given quantum state is separable or inseparable. The most prominent separability test is the Peres Horodecki Positivity of Partial Transpose (PPT) criterion [88, 94]. According to this criterion, if a state is separable, then the partial transposed density matrix is a positive matrix. Another similar criterion is proposed in ref. [95], which states that for a state to be separable, the determinant of the partial transposed density matrix should be positive. The above discussion shows the importance of partial transpose in the field of quantum entanglement studies. Let us take an example of a bipartite composite system in the Hilbert space $H = H_A \otimes H_B$ and the density matrix corresponding to the composite system is ρ_{AB}. We can write the matrix elements of ρ_{AB} as

$$\rho_{AB} = \sum_{ijkl} r_{ik;jl}|i,j\rangle\langle k,l| \equiv \sum_{ijkl} r_{ik;jl}(|i\rangle\langle k|)_A \otimes (|j\rangle\langle l|)_B. \tag{5.5}$$

In the above equation, indices i, k are basis states in the Hilbert space H_A of subsystem A and j, l are basis states in the Hilbert space H_B of subsystem B. Here $r_{ik;jl}$ are the matrix elements. The partial transpose of ρ_{AB} with respect to the system B is given by,

$$\rho_{AB}^{T_B} = \sum_{ijkl} r_{ik;jl} (|i\rangle\langle k|)_A \otimes (|l\rangle\langle j|)_B. \tag{5.6}$$

Of course, $\rho_{AB}^{T_A}$ also can be found easily using the above. For a more in-depth understanding regarding the mathematical operation of partial transpose, we have defined a two-qubit density matrix as

$$\rho_{AB} = \begin{pmatrix} a_{11} & a_{12} & a_{13} & a_{14} \\ a_{21} & a_{22} & a_{23} & a_{24} \\ a_{31} & a_{32} & a_{33} & a_{34} \\ a_{41} & a_{42} & a_{43} & a_{44} \end{pmatrix}. \tag{5.7}$$

After partially transposing the subsystem B of the density matrix in Eq. [5.7], we finally obtain the following,

$$\rho_{AB}^{T_B} = \begin{pmatrix} a_{11} & a_{21} & a_{13} & a_{23} \\ a_{12} & a_{22} & a_{14} & a_{24} \\ a_{13} & a_{41} & a_{33} & a_{43} \\ a_{32} & a_{42} & a_{34} & a_{44} \end{pmatrix}. \tag{5.8}$$

The recipes we have given can do partial transpose for any random partition of a bipartite multi-qubit system.

5.2.1   For a real or complex state

`ptranspose_vec(state,sub_tr)`

Parameters

In/Out	Argument	Description
[in]	state	It is a real or complex array of dimension $(0{:}2^N{-}1)$, which is the input density matrix
[in]	sub_tr	It is list of numbers designating the particular subsystems to be partially transposed. The list can store any natural number between 1 and N
[out]	denc2	denc2 is a real or complex array of dimension $(0{:}2^N{-}1,0{:}2^N{-}1)$, which is the required partially transposed matrix

Implementation

```
def ptranspose_vec(self,state,sub_tr):
    """
    Partial transpose operation on a quantum state
    Parameters
        state : It is a real or complex state.
        sub_tr : List of number designating the subsystems
                 to be partially transposed.
    Returns
        denc2: It is partially transposed density matrix


    """
    # Total number of qubits N
    N=int(math.log2(state.shape[0]))
    # Checking whether entered values of subsystems are correct or not
    assert set(sub_tr).issubset(set(np.arange(1,N+1))),\
    "Invalid subsystems to be traced out"
    # The data type of the state, real or complex
    typestate=str(state.dtype)
    # Checking whether state is complex or not
    if re.findall("^complex",typestate):
        # It will store transposed density matrix $\rho_{AB}^{T_B}$
        denc2=np.zeros([2**N,2**N],dtype=np.complex_)
        for i in range(state.shape[0]):
            vec_row=qobj.decimal_binary(i,N)
            for j in range(state.shape[0]):
                vec_col=qobj.decimal_binary(j,N)
                vec_row2=vec_row.copy()
                for k in sub_tr:
                    temp=vec_row2[k-1]
                    vec_row2[k-1]=vec_col[k-1]
                    vec_col[k-1]=temp
                row=qobj.binary_decimal(vec_row2)
                col=qobj.binary_decimal(vec_col)
                denc2[row,col]=state[i]*np.conjugate(state[j])
    else:
        # It will store transposed density matrix $\rho_{AB}^{T_B}$
        denc2=np.zeros([2**N,2**N],dtype='float64')
        for i in range(state.shape[0]):
            vec_row=qobj.decimal_binary(i,N)
            for j in range(state.shape[0]):
                vec_col=qobj.decimal_binary(j,N)
                vec_row2=vec_row.copy()
                for k in sub_tr:
                    temp=vec_row2[k-1]
                    vec_row2[k-1]=vec_col[k-1]
                    vec_col[k-1]=temp
                row=qobj.binary_decimal(vec_row2)
                col=qobj.binary_decimal(vec_col)
                denc2[row,col]=state[i]*state[j]
    return(denc2)
```

Example

In this example, we are performing the partial transpose over the second qubit of the two-qubit pure real state $|\psi_2\rangle$ and we will get ρ^{T_2} as the output.

```
from QuantumInformation import QuantumMechanics as QM
qobj=QM()
state2=np.zeros([2**2],dtype='float64')
for i in range(0,state2.shape[0]):
    state2[i]=i+1
state2=qobj.normalization_vec(state2)
sub_tr=[2]
denc2=pobj.ptranspose_vec(state2,sub_tr)
print(denc2)
```

The preceding codes prints the required partially transposed matrix as shown below.

```
[[0.03333333 0.06666667 0.1        0.2       ]
 [0.06666667 0.13333333 0.13333333 0.26666667]
 [0.1        0.13333333 0.3        0.4       ]
 [0.2        0.26666667 0.4        0.53333333]]
```

Similarly, we perform the partial transpose over the second qubit of the two-qubit pure complex state $|\psi_3\rangle$ and we will get ρ^{T_2} as the output.

```
from QuantumInformation import QuantumMechanics as QM
qobj=QM()
state2=np.zeros([2**2],dtype=np.complex_)
for i in range(0,state2.shape[0]):
    state2[i]=complex(i,i+1)
state2=qobj.normalization_vec(state2)
sub_tr=[2]
denc2=pobj.ptranspose_vec(state2,sub_tr)
file2 = open('denc2_mat.txt', 'w')
# writing matrix elements in the file
for i in range(0,denc2.shape[0]):
    for j in range(0,denc2.shape[1]):
        file2.write(str(denc2[i,j]))
        # to break the line between row elements of matrix
        file2.write('\n')
# Closing file
file2.close()
```

The file "denc2_mat.txt" contains the entries of required partially transposed matrix.

```
(0.022727272727272728+0j)
(0.045454545454545456-0.022727272727272728j)
(0.06818181818181819+0.045454545454545456j)
(0.18181818181818182+0.02272727272727272j)
(0.045454545454545456+0.022727272727272728j)
(0.11363636363636365+0j)
(0.09090909090909091+0.06818181818181819j)
(0.25+0.04545454545454547j)
(0.06818181818181819-0.045454545454545456j)
```

```
(0.09090909090909091-0.06818181818181819j)
(0.29545454545454547+0j)
(0.40909090909090917-0.022727272727272735j)
(0.18181818181818182-0.022727272727272772j)
(0.25-0.04545454545454547j)
(0.40909090909090917+0.022727272727272735j)
(0.5681818181818182+0j)
```

5.2.2 For a real or complex density matrix

ptranspose_den(denc,sub_tr)

Parameters

In/Out	Argument	Description
[in]	denc	It is a real or complex array of dimension $(0{:}2^N-1,0{:}2^N-1)$, which is the input density matrix
[in]	sub_tr	It is a list of numbers designating the particular subsystems to be partially transposed. This list can store any natural number between 1 and N
[out]	denc2	denc2 is a real or complex array of dimension $(0{:}2^N-1,0{:}2^N-1)$, which is the required partially transposed matrix

Implementation

```python
def ptranspose_den(self,denc,sub_tr):
    """
    Partial transpose operation on density matrix
    Parameters
        denc : It is a real or complex density matrix.
        sub_tr : List of number designating the subsystems
                to be partially transposed.
    Returns
        denc2: It is partially transposed density matrix
    """
    # Total number of qubits N
    N=int(math.log2(denc.shape[0]))
    # Checking whether entered values of subsystems are correct or not
    assert set(sub_tr).issubset(set(np.arange(1,N+1))),\
    "Invalid subsystems to be traced out"
    # The type density matrix, real or complex
    typestate=str(denc.dtype)
    # Checking density is complex
    if re.findall("^complex",typestate):
```

```
        # It will store transposed density matrix ρ_{AB}^{T_B}
        denc2=np.zeros([2**N,2**N],dtype=np.complex_)
        for i in range(denc.shape[0]):
            vec_row=qobj.decimal_binary(i,N)
            for j in range(denc.shape[1]):
                vec_col=qobj.decimal_binary(j,N)
                vec_row2=vec_row.copy()
                for k in sub_tr:
                    temp=vec_row2[k-1]
                    vec_row2[k-1]=vec_col[k-1]
                    vec_col[k-1]=temp
                row=qobj.binary_decimal(vec_row2)
                col=qobj.binary_decimal(vec_col)
                denc2[row,col]=denc[i,j]
    else:
        # It will store transposed density matrix ρ_{AB}^{T_B}
        denc2=np.zeros([2**N,2**N],dtype='float64')
        for i in range(denc.shape[0]):
            vec_row=qobj.decimal_binary(i,N)
            for j in range(denc.shape[1]):
                vec_col=qobj.decimal_binary(j,N)
                vec_row2=vec_row.copy()
                for k in sub_tr:
                    temp=vec_row2[k-1]
                    vec_row2[k-1]=vec_col[k-1]
                    vec_col[k-1]=temp
                row=qobj.binary_decimal(vec_row2)
                col=qobj.binary_decimal(vec_col)
                denc2[row,col]=denc[i,j]
    return(denc2)
```

Example

In this example, we are performing the partial transpose over the second qubit of the two-qubit real density matrix $|\psi_2\rangle\langle\psi_2|$ and we will get ρ^{T_2} as the output.

```
from QuantumInformation import QuantumMechanics as QM
qobj=QM()
state2=np.zeros([2**2],dtype='float64')
for i in range(0,state2.shape[0]):
    state2[i]=i+1
state2=qobj.normalization_vec(state2)
state2=qobj.outer_product_rvec(state2,state2)
sub_tr=[2]
denc2=pobj.ptranspose_den(state2,sub_tr)
print(denc2)
```

The preceding code prints the required partially transposed matrix as shown below.

```
[[0.03333333 0.06666667 0.1        0.2       ]
 [0.06666667 0.13333333 0.13333333 0.26666667]
 [0.1        0.13333333 0.3        0.4       ]
 [0.2        0.26666667 0.4        0.53333333]]
```

Similarly, we perform the partial transpose over the second qubit of the two-qubit complex density matrix $|\psi_3\rangle\langle\psi_3|$ and we will get ρ^{T_2} as the output.

```
from QuantumInformation import QuantumMechanics as QM
qobj=QM()
state2=np.zeros([2**2],dtype=np.complex_)
for i in range(0,state2.shape[0]):
    state2[i]=complex(i,i+1)
state2=qobj.normalization_vec(state2)
state2=qobj.outer_product_cvec(state2,state2)
sub_tr=[2]
denc2=pobj.ptranspose_den(state2,sub_tr)
file2 = open('denc2_mat.txt', 'w')
# writing matrix elements in the file
for i in range(0,denc2.shape[0]):
    for j in range(0,denc2.shape[1]):
        file2.write(str(denc2[i,j]))
        # to break the line between row elements of matrix
        file2.write('\n')
# Closing file
file2.close()
```

The file "denc2_mat.txt" contains the row entries of the required partially transposed matrix.

```
(0.022727272727272728+0j)
(0.045454545454545456-0.022727272727272728j)
(0.06818181818181819+0.045454545454545456j)
(0.18181818181818182+0.02272727272727272j)
(0.045454545454545456+0.022727272727272728j)
(0.11363636363636365+0j)
(0.09090909090909091+0.06818181818181819j)
(0.25+0.04545454545454547j)
(0.06818181818181819-0.045454545454545456j)
(0.09090909090909091-0.06818181818181819j)
(0.29545454545454547+0j)
(0.40909090909090917-0.022727272727272735j)
(0.18181818181818182-0.02272727272727272j)
(0.25-0.04545454545454547j)
(0.40909090909090917+0.022727272727272735j)
(0.5681818181818182+0j)
```

5.3 Concurrence

Concurrence [96, 97] is a measure of entanglement defined for both pure and mixed states of two-qubit. Consider a N-qubit quantum system and the quantum state is ρ_N. The pictorial presentation of such a system is shown in Fig. (5.3). To calculate the entanglement content between the ith and j^{th} qubits of the quantum state, we first calculate the two-qubit reduced density matrix $\rho_{ij} = Tr_{\overline{ij}}(\rho_N)$. Thereafter using the reduced density matrix ρ_{ij}, we calculate

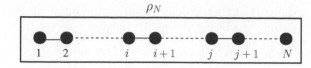

FIGURE 5.3: N-qubit quantum system, and the quantum state is denoted by ρ_N.

the concurrence value (two-qubit entanglement content) by the following mathematical expression:

$$C_{ij} = \text{Max}\left(0, \sqrt{\lambda_1} - \sqrt{\lambda_2} - \sqrt{\lambda_3} - \sqrt{\lambda_4}\right). \tag{5.9}$$

Here λ_i's are the eigenvalues of the matrix $\rho_{ij}\tilde{\rho}_{ij}$ in the non-increasing order. The matrix $\tilde{\rho}_{ij}$ is defined as,

$$\tilde{\rho}_{ij} = (\sigma_y \otimes \sigma_y)\rho_{ij}^*(\sigma_y \otimes \sigma_y). \tag{5.10}$$

Note carefully that if we have any pure or mixed state of N qubits and once we have the reduced density matrix between any two qubits i and j as ρ_{ij}, we can directly use the above prescription to calculate concurrence. It is worth mentioning that it is a measure of entanglement that has a closed-form expression for two-qubit states and at the same time cannot be generalized easily to higher dimensions. Note that in the recipes given, we can calculate the concurrence of any pair ij $(i < j)$ of a N-qubit state or a density matrix.

5.3.1   For a real or complex state

```
concurrence_vec(state,i,j,eps=10**(-13))
```

Parameters

In/Out	Argument	Description
[in]	state	It is a real or complex array of dimension $(0{:}2^N{-}1)$, which is the input state
[in]	i, j	It stores the place values of the qubits. Both of them can store any integer value between 1 and N
[in]	eps	If the magnitude of any of the eigenvalues of $\rho_{ij}\tilde{\rho}_{ij}$ is less than eps, then that particular value will be considered equal to zero. The default value is equal to 10^{-13}
[out]	conc	It returns the value of concurrence between qubits i and j specified above

Implementation

```
def concurrence_vec(self,state,i,j,eps=10**(-13)):
    """

    Calculation of concurrence for a quantum state
    Parameters
```

```
    state : Real or complex state
    i : It stores the place values of the qubits.
    j : It stores the place values of the qubits.
    eps : Below the eps value the eigenvalues will be considered zero.
          The default is 10**(-13).

Returns
    conc: concurrence value
"""
sigmay=np.zeros([4,4],dtype='float64')
# Storing the datatype of the input state
typestate=str(state.dtype)
# If data type of input state is complex then IF cond. is true
if re.findall("^complex",typestate):
    # Constructing σʸ ⊗ σʸ
    sigmay[0,3]=-1
    sigmay[1,2]=1
    sigmay[2,1]=1
    sigmay[3,0]=-1
    sub_tr=[i,j]
    # Constructing the two-qubit reduced density matrix ρᵢⱼ
    rdm= self.partial_trace_vec(state,sub_tr)
    # Calculating ρᵢⱼρ̃ᵢⱼ
    rhot3=rdm@sigmay@np.conjugate(rdm)@sigmay
    # Diagonalizing ρᵢⱼρ̃ᵢⱼ
    w,vl,vr,info =la.zgeev(rhot3)
    # wc will store eigenvalues of ρᵢⱼρ̃ᵢⱼ
    wc=[]
    for i in range(0,4):
        if abs(w.item(i))<eps:
            wc.append(0.000000000000000)
        else:
            wc.append(abs(w.item(i)))
    wc.sort(reverse=True)
    wc=np.array(wc,dtype='float64')
    # Calculating, √λ₁ - √λ₂ - √λ₃ - √λ₄
    conc=math.sqrt(wc.item(0))-math.sqrt(wc.item(1))-\
    math.sqrt(wc.item(2))-math.sqrt(wc.item(3))
    if conc<0:
        conc=0
# If data type of input state is real then ELSE cond. is true
else:
    # Constructing σʸ ⊗ σʸ
    sigmay[0,3]=-1
    sigmay[1,2]=1
    sigmay[2,1]=1
    sigmay[3,0]=-1
    sub_tr=[i,j]
    # Constructing two-qubit reduced density matrix ρᵢⱼ
    rdm= self.partial_trace_vec(state,sub_tr)
    # Calculating ρᵢⱼρ̃ᵢⱼ
```

```
        rhot3=rdm@sigmay@rdm@sigmay
        # Diagonalizing ρ_ij ρ̃_ij
        wr,wi,vl,vr,info =la.dgeev(rhot3)
        # w will be storing λ_i's
        w=[]
        for i in range(0,4):
            if wr[i] < eps:
                w.append(0.000000000000000)
            else:
                w.append(np.float64(wr.item(i)))
        w.sort(reverse=True)
        w=np.array(w,dtype='float64')
        # Finally calculating √λ_1 − √λ_2 − √λ_3 − √λ_4
        conc=math.sqrt(w.item(0))-math.sqrt(w.item(1))-\
        math.sqrt(w.item(2))-math.sqrt(w.item(3))
        if conc<0:
            conc=0.0
    return(np.float64(conc))
```

Example

In this example, we are finding the value of concurrence C_{35} of a six-qubit real state $|\psi_2\rangle$.

```
from QuantumInformation import QuantumMechanics as QM
qobj=QM()
# state2 will be storing |ψ_2⟩
state2=np.zeros([2**6],dtype='float64')
for i in range(0,state2.shape[0]):
    state2[i]=i+1
state2=qobj.normalization_vec(state2)
# Calculating the concurrence C_35
conc=entobj.concurrence_vec(state2,3,5)
print(conc)
```

The preceding code prints the required value of concurrence C_{35}.

```
0.0057245080501143785
```

In this example, we are finding the value of concurrence C_{36} of the six-qubit complex state $|\psi_3\rangle$.

```
from QuantumInformation import QuantumMechanics as QM
qobj=QM()
# state2 will be storing |ψ_3⟩
state2=np.zeros([2**6],dtype=np.complex_)
for i in range(0,state2.shape[0]):
    state2[i]=complex(i,i+1)
state2=qobj.normalization_vec(state2)
# Calculating the concurrence C_36
conc=entobj.concurrence_vec(state2,3,6)
print(conc)
```

The preceding code prints the required value of concurrence C_{36}.

0.002929329915577223

5.3.2   For a real or complex density matrix

```
concurrence_den(state,i,j,eps=10**(-13))
```

Parameters

In/Out	Argument	Description
[in]	state	It is a real or complex array of dimension $(0{:}2^N-1,0{:}2^N-1)$, which is the input density matrix
[in]	i, j	It stores the place values of the qubits. Both of them can store any integer value between 1 and N
[in]	eps	If the magnitude of any of the eigenvalues of $\rho_{ij}\tilde{\rho}_{ij}$ is less than eps, then that particular value will be considered equal to zero. The default value is equal to 10^{-13}
[out]	conc	It returns the value of concurrence between qubits i and j specified above

Implementation

```python
def concurrence_den(self,state,i,j,eps=10**(-13)):
    """
    Calculation of concurrence for a density matrix
    Parameters
        state : Real or complex density matrix
        i : It stores the place values of the qubits.
        j : It stores the place values of the qubits.
        eps : Below the eps value the eigenvalues will be considered zero.
            The default is 10**(-13).

    Returns
        conc: concurrence value
    """
    sigmay=np.zeros([4,4],dtype='float64')
    # Storing the data type of the input density matrix ρ
    typestate=str(state.dtype)
    # If ρ is complex then IF cond. is true
    if re.findall("^complex",typestate):
        # Constructing the σʸ ⊗ σʸ
        sigmay[0,3]=-1
        sigmay[1,2]=1
        sigmay[2,1]=1
        sigmay[3,0]=-1
```

```python
        sub_tr=[i,j]
        # Constructing the two-qubit reduced density matrix ρ_ij
        rdm= self.partial_trace_den(state,sub_tr)
        # Constructing ρ_ij ρ̃_ij
        rhot3=rdm@sigmay@np.conjugate(rdm)@sigmay
        w,vl,vr,info =la.zgeev(rhot3)
        # wc stores the eigenvalues λ_i's
        wc=[]
        for i in range(0,4):
            if abs(w.item(i))<eps:
                wc.append(0.000000000000000)
            else:
                wc.append(abs(w.item(i)))
        wc.sort(reverse=True)
        wc=np.array(wc,dtype='float64')
        # Finally, √λ_1 - √λ_2 - √λ_3 - √λ_4
        conc=math.sqrt(wc.item(0))-math.sqrt(wc.item(1))-\
        math.sqrt(wc.item(2))-math.sqrt(wc.item(3))
        if conc<0:
            conc=0
    # If ρ is real then ELSE cond. is true
    else:
        # Constructing σ^y ⊗ σ^y
        sigmay[0,3]=-1
        sigmay[1,2]=1
        sigmay[2,1]=1
        sigmay[3,0]=-1
        sub_tr=[i,j]
        # Constructing two-qubit reduced density matrix ρ_ij
        rdm= self.partial_trace_den(state,sub_tr)
        # Constructing matrix ρ_ij ρ̃_ij
        rhot3=rdm@sigmay@rdm@sigmay
        # Diagonalizing ρ_ij ρ̃_ij
        wr,wi,vl,vr,info =la.dgeev(rhot3)
        # w will store the λ_i's
        w=[]
        for i in range(0,4):
            if wr[i] < eps:
                w.append(0.000000000000000)
            else:
                w.append(np.float64(wr.item(i)))
        w.sort(reverse=True)
        w=np.array(w,dtype='float64')
        # Finally, constructing √λ_1 - √λ_2 - √λ_3 - √λ_4
        conc=math.sqrt(w.item(0))-math.sqrt(w.item(1))-\
        math.sqrt(w.item(2))-math.sqrt(w.item(3))
        if conc<0:
            conc=0.0
    return(np.float64(conc))
```

Example

In this example we are finding the value of the concurrence C_{36} of a six-qubit density matrix $|\psi_2\rangle\langle\psi_2|$.

```
from QuantumInformation import QuantumMechanics as QM
qobj=QM()

# state2 will store |ψ₂⟩⟨ψ₂|
state2=np.zeros([2**6],dtype='float64')
for i in range(0,state2.shape[0]):
    state2[i]=i+1
state2=qobj.normalization_vec(state2)
state2=qobj.outer_product_rvec(state2,state2)
# Finally calculating concurrence C₃₆
conc=entobj.concurrence_den(state2,3,6)
print(conc)
```

The preceding code prints the required value of concurrence C_{36}.

```
0.002862254025114736
```

Similarly, we also find the value of the concurrence C_{36} of the six-qubit density matrix $|\psi_3\rangle\langle\psi_3|$.

```
from QuantumInformation import QuantumMechanics as QM
qobj=QM()
# state2 will store |ψ₃⟩⟨ψ₃|
state2=np.zeros([2**6],dtype=np.complex_)
for i in range(0,state2.shape[0]):
    state2[i]=complex(i,i+1)
state2=qobj.normalization_vec(state2)
state2=qobj.outer_product_cvec(state2,state2)
# Finally calculating concurrence C₃₆
conc=entobj.concurrence_den(state2,3,6)
print(conc)
```

The preceding code prints the required value of concurrence C_{36}.

```
0.002929329915577223
```

In this final example, we calculate the value of concurrence C_{12} for the following mixed state,

$$\rho = \frac{1}{2}|\psi_2\rangle\langle\psi_2| + \frac{1}{2}|\psi_3\rangle\langle\psi_3| \tag{5.11}$$

```
from QuantumInformation import QuantumMechanics as QM
qobj=QM()

# Constructing the |ψ₂⟩⟨ψ₂|
state2=np.zeros([2**6],dtype='float64')
for i in range(0,state2.shape[0]):
    state2[i]=i+1
state2=qobj.normalization_vec(state2)
```

```
state2=qobj.outer_product_rvec(state2,state2)

# Constructing the |ψ₃⟩⟨ψ₃|
state3=np.zeros([2**6],dtype=np.complex_)
for i in range(0,state3.shape[0]):
    state3[i]=complex(i,i+1)
state3=qobj.normalization_vec(state3)
state3=qobj.outer_product_cvec(state3,state3)

# Finally, constructing the mixed state ρ
mix_state=(0.5*state2)+(0.5*state3)

# Calculating the concurrence C₁₂
conc=entobj.concurrence_den(mix_state,1,2)

print(conc)
```

The preceding code prints the required value of concurrence C_{12}.

```
0.1853306861064475
```

5.3.3   Variants of concurrence

There are different variants of concurrence in the entanglement literature as follows. *I*-concurrence [98] of any multi-qubit pure state is defined as:

$$C_I = \sqrt{2(1 - Tr(\rho_A^2))},\tag{5.12}$$

where ρ_A is the reduced density matrix of partition A. The example of finding the *I*-concurrence for state $(|00\rangle + |11\rangle)/\sqrt{2}$ is given below,

```
import numpy as np
from QuantumInformation import PartialTr as PT
from QuantumInformation import GatesTools as GT

pobj=PT()
gtobj=GT()

# Created the state, (|00⟩ + |11⟩)/√2
state=gtobj.bell3(2)

# Calculated the reduced density matrix, ρ_A
red_A=pobj.partial_trace_vec(state,[2])

I_conc=np.sqrt(2*(1-np.trace(np.matmul(red_A,\
                                    red_A))))

# Finally, calculating C_I = √(2(1 - Tr(ρ²_A))
print(f"The I-Concurrence value is = {I_conc}")
```

The preceding code generates the following output.

The I-Concurrence value is = 1.0000000000000002

There exists another type of concurrence called the *N*-concurrence [99] which is defined as,

$$C_N = 2^{1-\frac{N}{2}} \sqrt{2^N - 2 - \sum_i Tr(\rho_i^2)}. \tag{5.13}$$

The example of finding the *N*-concurrence for state $(|00\rangle + |11\rangle)/\sqrt{2}$ is given below,

```
import numpy as np
from QuantumInformation import PartialTr as PT
from QuantumInformation import GatesTools as GT

pobj=PT()
gtobj=GT()

N=2
# Created the state, (|00⟩ + |11⟩)/√2
state=gtobj.bell3(N)

sumtr=0.0
# Calculating the value, ∑ᵢTr(ρᵢ²)
for i in range(1,N+1):
    red_i = pobj.partial_trace_vec(state,[i])
    sumtr = sumtr+np.trace(np.matmul(red_i,red_i))

# Finally, calculating the Cₙ = 2¹⁻ᴺ/² √2ᴺ-2-∑ᵢTr(ρᵢ²).
N_conc = 2**(1-(N/2))*np.sqrt(2**N-2-sumtr)

print(f"The N-Concurrence value is = {N_conc}")
```

The preceding code generates the following output.

The N-Concurrence value is = 1.0000000000000002

5.4 Block entropy

Consider a two-qubit pure state density matrix ρ^{AB}, where A and B are the two subsystems. The von Neumann [24, 100] entropy for the pure state ρ^{AB} is defined as follows,

$$S\left(\rho^{AB}\right) = -Tr\left(\rho^A \log_2 \rho^A\right) = -Tr\left(\rho^B \log_2 \rho^B\right), \tag{5.14}$$

where ρ^A and ρ^B are the reduced density matrices of the subsystems A and B, respectively. Eq. [5.14] holds true because of Schmidt decomposition [101], which states that non-zero eigenvalues of the reduced density matrix ρ^A are equal to non-zero eigenvalues of reduced density matrix ρ^B, if the composite state of the bipartite split is a pure state. Eq. [5.14] can also be written as,

$$S\left(\rho^{AB}\right) = -\sum_{i=1}^{2} \lambda_i \log_2\left(\lambda_i\right), \tag{5.15}$$

where λ_i's are the eigenvalues of the reduced density matrix ρ^B (or ρ^A), and we define $0 \log_2 0 \equiv 0$. The entanglement content of a pure state is expressed in terms of the reduced density matrices of the subsystems. The loss of information or ignorance in the state of the subsystem even though the whole system may be completely known is quantified in terms of entropy.

For a multi-qubit pure state, the entanglement between two blocks of spins is calculated by block entropy [102]. In Fig. (5.4), we have N spins of a multi-qubit pure state and we calculate the entanglement between L ($L < N$) and $N - L$ block of spins. The block entropy gives the entanglement between these two blocks of spins as follows,

$$S(\rho_L) = -\sum_{i=1}^{d_L} \lambda_i \log_2 \lambda_i. \tag{5.16}$$

where λ_i are the eigenvalues of the reduced density matrix $\rho_L = tr_{L+1,...,N}(\rho)$, and d_L is the Hilbert space dimension of the reduced density matrix ρ_L. Note that Eq. [5.16] is a concave function. Now we will find the upper bound of the block entropy, and to this end, we rewrite the Eq. [5.16] as follows by absorbing the negative sign inside,

$$S(\rho_L) = \sum_{i=1}^{d_L} \lambda_i \log_2 \frac{1}{\lambda_i}. \tag{5.17}$$

By using the definition of concave function, the above equation transform into the following

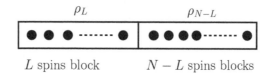

ρ_L $\qquad$ ρ_{N-L}

L spins block $\qquad$ $N - L$ spins blocks

FIGURE 5.4: Entanglement between L and $N - L$ blocks of spins.

inequality,

$$S(\rho_L) = \sum_{i=1}^{d_L} \lambda_i \log_2 \frac{1}{\lambda_i} \leq \log_2 \left(\sum_{i=1}^{d_L} \lambda_i \frac{1}{\lambda_i} \right), \tag{5.18}$$

$$S(\rho_L) \leq \log_2(d_L). \tag{5.19}$$

The equality in Eq. [5.19] holds true only in the case when all λ_i's are equal to $1/d_L$, and it is only possible when the state is a maximally mixed state, $\mathbb{I}/d_L$. The upper and lower bound of block entropy can be simply written as,

$$0 \leq S(\rho_L) \leq log_2(d_L). \tag{5.20}$$

$S(\rho_L)$ is equal to zero when ρ_L is a pure state (where all the λ_i's are zero except any one of the λ_i value which is equal to 1).

5.4.1 For a real or complex state

```
block_entropy_vec(state,sub_tr,eps=10**(-13))
```

Parameters

In/Out	Argument	Description
[in]	state	It is a real or complex array of dimension $(0{:}2^N-1)$, which is the input state
[in]	sub_tr	It is a list of numbers designating the particular subsystems not to be traced out. The list can store any natural number between 1 and N
[in]	eps	If the magnitude of any of the eigenvalues of ρ_A is less than eps, then that particular value will be considered equal to zero. The default value is equal to 10^{-13}
[out]	Bent	It returns the value of block entropy

Implementation

```python
def block_entropy_vec(self,state,sub_tr,eps=10**(-13)):
    """
    Calculation of block entropy for a quantum state
    Parameters
        state : Real or complex state
        sub_tr: List of numbers designating the particular subsystems
                not to be traced out.
        eps : Below the eps value the eigenvalues will be considered zero.
              The default is 10**(-13).

    Returns
        Bent: Block entropy value
    """
    # Storing the data type of the state |ψ⟩
    typestate=str(state.dtype)
    # Calculating the reduced density matrix ρ_L
    rdm= self.partial_trace_vec(state,sub_tr)
    # If |ψ⟩ is complex then IF cond. is true
    if re.findall("^complex",typestate):
        # Diagonalizing ρ_L
        w,v,info=la.zheev(rdm)
    # If |ψ⟩ is real then ELSE cond. is true
    else:
        # Diagonalizing ρ_L
        w,v,info=la.dsyev(rdm)
    wlen=len(w)
    Bent=0.0
    # Finally calculating S(ρ_L) = -∑_{i=1}^{d_L} λ_i log_2 λ_i
    for x in range(0,wlen):
        if abs(w.item(x))<eps:
            w[x]=0.000000000000000
        else:
```

```
        assert w.item(x) > 0.0,\
        "The density matrix entered is not correct as the eigenvalues
    are negative"
        Bent=Bent-(w.item(x)*math.log(w.item(x),2))
    return(Bent)
```

Example

In this example, we are finding the value of the block entropy between the blocks 135 and 246 of the six-qubit real state $|\psi_2\rangle$.

```
from QuantumInformation import QuantumMechanics as QM
qobj=QM()
# state2 will be storing |ψ₂⟩
state2=np.zeros([2**6],dtype='float64')
for i in range(0,state2.shape[0]):
    state2[i]=i+1
state2=qobj.normalization_vec(state2)
sub_tr=[1,3,5]
# Calculating block entropy S(ρ₁₃₅)
bent=entobj.block_entropy_vec(state2,sub_tr)
print(bent)
```

The preceding code prints the required value of block entropy.

```
0.07835081818130613
```

Similarly, we find the value of block entropy between the blocks 135 and 246 of the six-qubit complex state $|\psi_3\rangle$.

```
from QuantumInformation import QuantumMechanics as QM
qobj=QM()
# state2 will be storing |ψ₃⟩
state2=np.zeros([2**6],dtype=np.complex_)
for i in range(0,state2.shape[0]):
    state2[i]=complex(i,i+1)
state2=qobj.normalization_vec(state2)
sub_tr=[1,3,5]
# Calculating block entropy S(ρ₁₃₅)
bent=entobj.block_entropy_vec(state2,sub_tr)
print(bent)
```

The preceding code prints the required value of block entropy.

```
0.08141948096557698
```

5.4.2   For a real or complex density matrix

```
block_entropy_den(state,sub_tr,eps=10**(-13))
```

Parameters

In/Out	Argument	Description
[in]	state	It is a real or complex array of dimension $(0:2^N-1,0:2^N-1)$, which is the input density matrix
[in]	sub_tr	It is a list of numbers designating the particular subsystems not to be traced out. The list can store any natural number between 1 and N
[in]	eps	If the magnitude of any of the eigenvalues of ρ_A is less than eps, then that particular value will be considered equal to zero. The default value is equal to 10^{-13}
[out]	Bent	It returns the value of block entropy

Implementation

```
def block_entropy_den(self,state,sub_tr,eps=10**(-13)):
    """
    Calculation of block entropy for a density matrix
    Parameters
        state : Real or complex density matrix
        sub_tr: List of numbers designating the particular subsystems
                not to be traced out.
        eps : Below the eps value the eigenvalues will be considered zero.
              The default is 10**(-13).

    Returns
        Bent: Block entropy value
    """
    # Storing the data type of input density matrix ρ
    typestate=str(state.dtype)
    # Calculating reduced density matrix ρ_L
    rdm= self.partial_trace_den(state,sub_tr)
    # If ρ is complex then IF cond. is true
    if re.findall("^complex",typestate):
        # Diagonalizing ρ_L
        w,v,info=la.zheev(rdm)
    # If ρ is real then ELSE cond. is true
    else:
        # Diagonalizing ρ_L
        w,v,info=la.dsyev(rdm)
    wlen=len(w)
    Bent=0.0
    # Finally calculating S(ρ_L) = - Σ_{i=1}^{d_L} λ_i log_2 λ_i
    for x in range(0,wlen):
        if abs(w.item(x))<eps:
```

```
        w[x]=0.000000000000000
    else:
        assert w.item(x) > 0.0,\
        "The density matrix entered is not correct as the eigenvalues
are negative"
        Bent=Bent-(w.item(x)*math.log(w.item(x),2))
    return(Bent)
```

Example

In this example, we are finding the value of block entropy between the blocks 135 and 246 of the six-qubit real density matrix $|\psi_2\rangle\langle\psi_2|$.

```
from QuantumInformation import QuantumMechanics as QM
qobj=QM()
# state2 will be storing |ψ₂⟩⟨ψ₂|
state2=np.zeros([2**6],dtype='float64')
for i in range(0,state2.shape[0]):
    state2[i]=i+1
state2=qobj.normalization_vec(state2)
state2=qobj.outer_product_rvec(state2,state2)
sub_tr=[2,4,6]
# Calculating block entropy S(ρ₂₄₆)
bent=entobj.block_entropy_den(state2,sub_tr)
print(bent)
```

The preceding code prints the required value of block entropy.

```
0.0783508181813065
```

Similarly, we also find the value of block entropy between the blocks 135 and 246 of the six-qubit density matrix $|\psi_3\rangle\langle\psi_3|$.

```
from QuantumInformation import QuantumMechanics as QM
qobj=QM()
# state2 will be storing |ψ₃⟩⟨ψ₃|
state2=np.zeros([2**6],dtype=np.complex_)
for i in range(0,state2.shape[0]):
    state2[i]=complex(i,i+1)
state2=qobj.normalization_vec(state2)
state2=qobj.outer_product_cvec(state2,state2)
sub_tr=[2,4,6]
# Calculating block entropy S(ρ₂₄₆)
bent=entobj.block_entropy_den(state2,sub_tr)
print(bent)
```

The preceding code prints the required value of block entropy.

```
0.08141948096557718
```

5.5 Renyi entropy

In classical information theory, the Renyi entropy [103, 104] generalizes the following entropies: Shannon entropy [60], collision entropy [105] and the min-entropy [106]. However, the quantum version of the Renyi entropy (S_α) quantifies the quantum entanglement content of a quantum system. It is defined using two quantities, the density matrix and Renyi index $\alpha \geq 0$. Renyi entropy is defined as follows,

$$S_\alpha = \frac{1}{1-\alpha} \log Tr(\rho^\alpha). \tag{5.21}$$

Here α varies from 0 to ∞ such that $\alpha \neq 1$, ρ is a density matrix or reduced density matrix describing the state of the system. In the limit $\alpha \to 1$, the Renyi entropy approaches the von Neumann entanglement entropy. Renyi entropy is used in many areas of quantum information theory.

5.5.1 For a real or complex density matrix

renyi_entropy(rho,alpha)

Parameters

In/Out	Argument	Description
[in]	rho	rho is a real or complex array of dimension ($0{:}2^N-1,0{:}2^N-1$), which is the input density matrix
[in]	alpha	It is the value of Renyi index
[out]	renyi	It returns the value of Renyi entropy

Implementation

```
def renyi_entropy(self,rho,alpha):
    """
    Calculation of Renyi entropy
    Parameters
        rho : Real or complex density matrix
        alpha : It is the value of Renyi index

    Returns
        renyi : Renyi Entropy value
    """
    # Checking whether α ≠ 1
    assert alpha != 1.0, "alpha should not be equal to 1"
    # Storing the data type of input density matrix ρ
    typerho=str(rho.dtype)
    # Creating the object of the LinearAlgebra class
    laobj=LA()
```

```
    # If ρ is complex then IF cond. is true
    if re.findall('^complex',typerho):
        # Finally calculating, S_α = 1/(1-α) log Tr(ρ^α)
        renyi=math.log(abs(np.trace(laobj.power_hmatrix(rho,\
        alpha))))/(1-alpha)
    # If ρ is real then ELSE cond. is true
    else:
        # Finally calculating, S_α = 1/(1-α) log Tr(ρ^α)
        renyi=math.log(np.trace(laobj.power_smatrix(rho,alpha)))/(1-alpha)
    return renyi
```

Example

In this example, we are finding the value of Renyi entropy of a two qubit generalized Werner state $\rho(0.5)$.

```
from QuantumInformation import GatesTools as GT
gtobj=GT()
# Constructing two-qubit Werner state with mixing probability p = 0.5
state=gtobj.nWerner(0.5)
# Finally calculating Renyi entropy
renyi=entobj.renyi_entropy(state,6.0)
print(renyi)
```

The preceding prints the required value of Renyi entropy.

```
0.5639659587808111
```

Similarly, we calculate the value of Renyi entropy of the density matrix $\rho_1 \otimes \rho_2$, where density matrices ρ_1 and ρ_2 are already defined in Eq. [4.8] and Eq. [4.9], respectively.

```
# Constructing the density matrix ρ_1 ⊗ ρ_2
rho=np.zeros([4,4],dtype=np.complex_)
rho[0,0]=complex(0.16,0.0)
rho[0,1]=complex(-0.04,-0.16)
rho[0,2]=complex(-0.08,-0.04)
rho[0,3]=complex(-0.02,0.09)
rho[1,0]=complex(-0.04,0.16)
rho[1,1]=complex(0.24,0.0)
rho[1,2]=complex(0.06,-0.07)
rho[1,3]=complex(-0.12,-0.06)
rho[2,0]=complex(-0.08,0.04)
rho[2,1]=complex(0.06,0.07)
rho[2,2]=complex(0.24,0.0)
rho[2,3]=complex(-0.06,-0.24)
rho[3,0]=complex(-0.02,-0.09)
rho[3,1]=complex(-0.12,0.06)
rho[3,2]=complex(-0.06,0.24)
rho[3,3]=complex(0.36,0.0)
# Finally calculating Renyi entropy
renyi=entobj.renyi_entropy(rho,100.0)
print(renyi)
```

The preceding code prints the required value of Renyi entropy.

```
0.37696668017247514
```

5.6 Negativity and logarithmic negativity

Negativity [107] is an entanglement measure based on the partial transpose operation. There is no operational interpretation of negativity. Negativity $\mathcal{N}(\rho)$ for a given bipartite pure or mixed state density matrix ρ is defined with respect to the partial transpose on one of its subsystems A as,

$$\mathcal{N}(\rho) = \frac{\| \rho^{T_A} \| - 1}{2}. \tag{5.22}$$

The mathematical construction of Eq. [5.22] is such that for separable states, the negativity value goes to zero, and in the case of maximally entangled states, its value is equal to entanglement entropy. It measures the degree to which ρ^{T_A} fails to be positive, and therefore it can be regarded as a quantitative version of Peres' criterion for separability (PPT criteria). Associated with negativity, we also define what is called the logarithmic negativity [108] which is $\mathcal{E}_N(\rho) = \log_2(\| \rho^{T_A} \|)$, as expected of a log function, logarithmic negativity has additive properties. Though logarithmic negativity is not a convex function but its value never increases on average under any general positive partial transpose preserving operation.

5.6.1 For a real or complex state

```
negativity_log_vec(state,sub_tr,eps=10**(-13))
```

Parameters

In/Out	Argument	Description
[in]	state	It is a real or complex array of dimension $(0{:}2^N-1)$, which is the input state
[in]	sub_tr	It is a list of numbers designating the particular subsystems to be transposed. The list can store any natural number between 1 and N
[in]	eps	If the magnitude of any of the eigenvalues of ρ^{T_A} is less than eps, then that particular value will be considered equal to zero. The default value is equal to 10^{-13}
[out]	negv, lognegv	Returns the value of negativity and logarithmic negativity

Implementation

```
def negativity_log_vec(self,state,sub_tr,eps=10**(-13)):
    """
    Calculation of negativity and logarithmic negativity for a quantum
    state
    Parameters
        state : Real or complex state
        sub_tr: List of numbers designating the particular subsystems
                to be transposed.
        eps : Below the eps value the eigenvalues will be considered zero.
              The default is 10**(-13).

    Returns
        negv,lognegv : negativity and log negativity values, respectively
    """
    # Constructing the object of the LinearAlgebra class
    laobj=LA()
    # Storing the data type of the input state, |ψ⟩
    typestate=str(state.dtype)
    # Calculated partially transposed matrix, ρ^{T_A}
    rhoa=self.ptranspose_vec(state,sub_tr)
    # If |ψ⟩ is complex then IF cond. is true
    if re.findall("^complex",typestate):
        # Calculating, 𝒩(ρ) = (‖ρ^{T_A}‖ − 1) / 2
        negv=laobj.trace_norm_cmatrix(rhoa,precision=eps)
    # If |ψ⟩ is real then ELSE cond. is true
    else:
        # Calculating, 𝒩(ρ) = (‖ρ^{T_A}‖ − 1) / 2
        negv=laobj.trace_norm_rmatrix(rhoa,precision=eps)
    # The value of 𝒩(ρ) should be greater than 0.
    assert negv > 0.0,\
    "The density matrix entered is not correct as the negativity is
    negative"
    # Finally calculating, ℰ_N(ρ) = log₂(‖ρ^{T_A}‖)
    lognegv=math.log2(negv)
    negv=(negv-1)/2
    return(negv,lognegv)
```

Example

In this example, we are finding the value of negativity and logarithmic negativity between the blocks 135 and 246 of the six-qubit real state $|\psi_2\rangle$.

```
from QuantumInformation import QuantumMechanics as QM
qobj=QM()
# state2 will store |ψ₂⟩
state2=np.zeros([2**6],dtype='float64')
for i in range(0,state2.shape[0]):
```

```
    state2[i]=i+1
state2=qobj.normalization_vec(state2)
# Designating the systems that are to be transposed
sub_tr=[1,3,5]
# Calculating the  negativity and logarithmic negativity
negv,logegv=entobj.negativity_log_vec(state2,sub_tr)
print(negv,logegv)
```

The preceding code prints the required values of negativity and logarithmic negativity.

```
0.0976744186046512 0.2574316996044179
```

Similarly, we calculate the value of negativity and logarithmic negativity between the blocks 135 and 246 of the six-qubit complex state $|\psi_3\rangle$.

```
from QuantumInformation import QuantumMechanics as QM
qobj=QM()
# state2 will store |ψ₃⟩
state2=np.zeros([2**6],dtype=np.complex_)
for i in range(0,state2.shape[0]):
    state2[i]=complex(i,i+1)
state2=qobj.normalization_vec(state2)
# Designating the systems that are to be transposed
sub_tr=[1,3,5]
# Calculating the  negativity and logarithmic negativity
negv,lognegv=entobj.negativity_log_vec(state2,sub_tr)
print(negv,lognegv)
```

The preceding code prints the required values of negativity and logarithmic negativity.

```
0.09996338337605282 0.26294635877747374
```

5.6.2   For a real or complex density matrix

```
negativity_log_den(den,sub_tr,eps=10**(-13))
```

Parameters

In/Out	Argument	Description
[in]	den	It is a real or complex array of dimension $(0{:}2^N{-}1,0{:}2^N{-}1)$, which is the input density matrix
[in]	sub_tr	It is list of numbers designating the particular subsystems to be transposed. The list can store any natural number between 1 and N
[in]	eps	If the magnitude of any of the eigenvalues of ρ^{T_A} is less than eps, then that particular value will be considered equal to zero. The default value is equal to 10^{-13}
[out]	negv, lognegv	Returns the value of negativity and logarithmic negativity

Implementation

```python
def negativity_log_den(self,den,sub_tr,eps=10**(-13)):
    """
    Calculation of negativity and logarithmic negativity for a density
    matrix
    Parameters
        state : Real or complex density matrix
        sub_tr: List of numbers designating the particular subsystems
                to be transposed.
        eps : Below the eps value the eigenvalues will be considered zero.
              The default is 10**(-13).

    Returns
        negv,lognegv : negativity and log negativity values, respectively
    """
    # Constructing the object of the LinearAlgebra class
    laobj=LA()
    # Storing the data type of the input state, ρ
    typestate=str(den.dtype)
    # Calculated partially transposed matrix, ρ^{T_A}
    rhoa=self.ptranspose_den(den,sub_tr)
    # If ρ is complex then IF cond. is true
    if re.findall("^complex",typestate):
        # Calculating, N(ρ) = (|| ρ^{T_A} || −1) / 2
        negv=laobj.trace_norm_cmatrix(rhoa,precision=eps)
    # If ρ is real then ELSE cond. is true
    else:
        # Calculating, N(ρ) = (|| ρ^{T_A} || −1) / 2
        negv=laobj.trace_norm_rmatrix(rhoa,precision=eps)
    # The value of N(ρ) should be greater than 0.
    assert negv > 0.0,\
    "The density matrix entered is not correct as the negativity is
    negative"
    # Finally calculating, E_N(ρ) = log_2(|| ρ^{T_A} ||)
    lognegv=math.log2(negv)
    negv=(negv-1)/2
    return(negv,lognegv)
```

Example

In this example, we are finding the value of negativity and logarithmic negativity between the block 135 and 246 of the six-qubit state $|\psi_2\rangle\langle\psi_2|$.

```python
from QuantumInformation import QuantumMechanics as QM
qobj=QM()
# state2 will store |ψ_2⟩
state2=np.zeros([2**6],dtype='float64')
for i in range(0,state2.shape[0]):
```

```
        state2[i]=i+1
state2=qobj.normalization_vec(state2)
# Constructing the density matrix |ψ2⟩⟨ψ2|
state2=qobj.outer_product_rvec(state2,state2)\
# Designating the subsystems that are to be transposed
sub_tr=[2,4,6]
# Calculating negativity and logarithmic negativity
negv, lognegv=entobj.negativity_log_den(state2,sub_tr)
print(negv, lognegv)
```

The preceding code prints the required values of negativity and logarithmic negativity.

```
0.0976744186046512 0.2574316996044179
```

Similarly, we calculate the value of negativity and logarithmic negativity between the blocks 135 and 246 of the six-qubit state $|\psi_3\rangle\langle\psi_3|$.

```
from QuantumInformation import QuantumMechanics as QM
qobj=QM()
# state2 will store |ψ3⟩
state2=np.zeros([2**6],dtype=np.complex_)
for i in range(0,state2.shape[0]):
    state2[i]=complex(i,i+1)
state2=qobj.normalization_vec(state2)
# Constructing the density matrix |ψ3⟩⟨ψ3|
state2=qobj.outer_product_cvec(state2,state2)
# Designating the subsystems that are to be transposed
sub_tr=[2,4,6]
# Calculating negativity and logarithmic negativity
negv, lognegv=entobj.negativity_log_den(state2,sub_tr)
print(negv, lognegv)
```

The preceding code prints the required values of negativity and logarithmic negativity.

```
0.09996338337605282 0.26294635877747374
```

In the final example, we construct the following mixed state density matrix,

$$\rho = \frac{1}{2}|\psi_2\rangle\langle\psi_2| + \frac{1}{2}|\psi_3\rangle\langle\psi_3|, \tag{5.23}$$

and we calculate the negativity and log-negativity values between the block 123 and 456.

```
from QuantumInformation import QuantumMechanics as QM
qobj=QM()

# state2 will store |ψ2⟩⟨ψ2|
state2=np.zeros([2**6],dtype='float64')
for i in range(0,state2.shape[0]):
    state2[i]=i+1
state2=qobj.normalization_vec(state2)
state2=qobj.outer_product_rvec(state2,state2)

# state2 will store |ψ3⟩⟨ψ3|
state3=np.zeros([2**6],dtype=np.complex_)
```

```
for i in range(0,state3.shape[0]):
    state3[i]=complex(i,i+1)
state3=qobj.normalization_vec(state3)
state3=qobj.outer_product_cvec(state3,state3)

# Finally, constructing the mixed state ρ
mix_state=(0.5*state2)+(0.5*state3)

# Designating the subsystems that are to be transposed
sub_tr=[4,5,6]

# Calculating the negativity and log-negativity
negv, log_negv = entobj.negativity_log_den(mix_state,sub_tr)

print(negv, log_negv)
```

The preceding code prints the required values of negativity and logarithmic negativity.

0.030392688074903806 0.08513279189272352

5.7 Q measure or the Meyer-Wallach-Brennen measure

Q measure [109, 110] is an entanglement measure for a multi-qubit pure state in the Hilbert space of dimension $(\mathbb{C}^2)^{\otimes N}$; it is defined as,

$$Q = 2 \left(1 - \frac{1}{N} \sum_{i=1}^{N} Tr(\rho_i^2) \right) \tag{5.24}$$

where ρ_i is a single qubit density matrix got by tracing all the other qubits except the ith qubit, and N is the total number of qubits. From the above expression it is clear that the Q measure is the average subsystem linear entropy of the constituent qubits. The bound on Q is such that, $0 \leq Q \leq 1$.

5.7.1 For a real or complex state

```
QMeasure_vec(state)
```

Parameters

In/Out	Argument	Description
[in]	state	It is a real or complex array of dimension $(0:2^N - 1)$, which is the input state
[out]	Qmeas	It returns the value of Q measure

Implementation

```
def QMeasure_vec(self,state):
    """
    Calculation of Q measure for a quantum state
    Parameters
        state : Real or complex state

    Returns
        Qmeas: Q measure value

    """
    # NN will store the total number of qubits
    NN=math.log2(state.shape[0])/math.log2(2)
    NN=int(NN)
    sub_tr=np.zeros([NN,1])
    sum3=0.0
    # FOR loop will store the value, \sum_{i=1}^{N} Tr(\rho_i^2)
    for x in range(0,NN):
        sub_tr=[]
        sub_tr.append(x+1)
        rho=self.partial_trace_vec(state,sub_tr)
        rho=np.matmul(rho,rho)
        tr2=np.trace(rho)
        sum3=sum3+tr2
    # Finally calculating, 2 ( 1 - 1/N \sum_{i=1}^{N} Tr(\rho_i^2) )
    Qmeas=2*(1-(sum3/NN))
    return abs(Qmeas)
```

Example

In this example, we are finding the value of Q measure of the six-qubit real state $|\psi_2\rangle$.

```
from QuantumInformation import QuantumMechanics as QM
qobj=QM()
# state2 will store |\psi_2\rangle
state2=np.zeros([2**6],dtype='float64')
for i in range(0,state2.shape[0]):
    state2[i]=i+1
state2=qobj.normalization_vec(state2)
# Calculating Q measure
Qmeas=entobj.QMeasure_vec(state2)
print(Qmeas)
```

The preceding code prints the required value of Q measure.

```
0.015888838041352438
```

Similarly, we find the value of Q measure of the six-qubit complex state $|\psi_3\rangle$.

```
from QuantumInformation import QuantumMechanics as QM
qobj=QM()
```

```
# state2 will store |ψ₃⟩
state2=np.zeros([2**6],dtype=np.complex_)
for i in range(0,state2.shape[0]):
    state2[i]=complex(i,i+1)
state2=qobj.normalization_vec(state2)
# Calculating Q measure
Qmeas=entobj.QMeasure_vec(state2)
print(Qmeas)
```

The preceding code prints the required value of Q measure.

```
0.01664226228792076
```

5.7.2   For a real or complex density matrix

```
QMeasure_den(den)
```

Parameters

In/Out	Argument	Description
[in]	den	It is a real or complex array of dimension $(0{:}2^N-1, 0{:}2^N-1)$ which is the input density matrix
[out]	Qmeas	It returns the value of Q measure

Implementation

```
def QMeasure_den(self,den):
    """

    Calculation of Q measure for a density matrix
    Parameters
        den : Real or complex density matrix

    Returns
        Qmeas: Q measure value
    """
    # NN stores total number of qubits
    NN=math.log2(den.shape[0])/math.log2(2)
    NN=int(NN)
    sub_tr=np.zeros([NN,1])
    sum3=0.0
    # FOR loop will store the value, ∑ᵢ₌₁ᴺ Tr(ρᵢ²)
    for x in range(0,NN):
        sub_tr=[]
        sub_tr.append(x+1)
        rho=self.partial_trace_den(den,sub_tr)
        rho=np.matmul(rho,rho)
        tr2=np.trace(rho)
```

```
        sum3=sum3+tr2
    # Finally calculating, 2(1 - 1/N Σ_{i=1}^{N} Tr(ρ_i²))
    Qmeas=2*(1-(sum3/NN))
    return abs(Qmeas)
```

Example

In this example, we are finding the value of Q measure of the six-qubit real density matrix $|\psi_2\rangle\langle\psi_2|$.

```
from QuantumInformation import QuantumMechanics as QM
qobj=QM()
# state2 will store |ψ₂⟩⟨ψ₂|
state2=np.zeros([2**6],dtype='float64')
for i in range(0,state2.shape[0]):
    state2[i]=i+1
state2=qobj.normalization_vec(state2)
state2=qobj.outer_product_rvec(state2,state2)
# Calculating the Q measure
Qmeas=entobj.QMeasure_den(state2)
print(Qmeas)
```

The preceding code prints the required value of Q measure.

```
0.015888838041352438
```

Similarly, we calculate the value of Q measure of the six-qubit complex density matrix $|\psi_3\rangle\langle\psi_3|$.

```
from QuantumInformation import QuantumMechanics as QM
qobj=QM()
# state2 will store |ψ₃⟩⟨ψ₃|
state2=np.zeros([2**6],dtype=np.complex_)
for i in range(0,state2.shape[0]):
    state2[i]=complex(i,i+1)
state2=qobj.normalization_vec(state2)
state2=qobj.outer_product_cvec(state2,state2)
# Calculating the Q measure
Qmeas=entobj.QMeasure_den(state2)
print(Qmeas)
```

The preceding code prints the required value of Q measure.

```
0.01664226228792076
```

5.8   Entanglement spectrum

The entanglement spectrum [111, 112] is the negative natural logarithm of the eigenvalue spectrum of the reduced density matrix. Instead of having a number (von Neumann entropy) for quantifying entanglement, Haldane in his seminal work pointed out that the spectrum of the reduced density matrix contains more information regarding the entanglement. This quantity is more important when we study condensed matter systems like topological materials, quantum Hall effect, etc. If we have a reduced density matrix of a composite system AB as ρ_A, then the entanglement spectrum is defined as $-ln(\lambda_i)$, where λ_i with $i = 1, \cdots, \dim(\rho_A)$ are the eigenvalues of ρ_A.

5.8.1   For a real or complex density matrix

---

entanglement_spectrum(rho)

Parameters

In/Out	Argument	Description
[in]	rho	It is a real or complex array of dimension $(0{:}2^N - 1, 0{:}2^N - 1)$, which is the input density matrix ρ_A
[out]	eigenvalues	List containing the eigenvalues of rho
[out]	logeigenvalues	List containing the negative logarithmic eigenvalues of rho, i.e. $-ln(\lambda_i)$ values

Implementation

```python
def entanglement_spectrum(self,rho):
    """
    Calculation of entanglement spectrum of a density matrix
    Parameters
        rho :  Real or complex density matrix

    Returns
        eigenvalues : List containing the eigenvalues of rho
        logeigenvalues : List containing the negative logarithmic
                         eigenvalues of rho
    """
    # Storing the data type of input state |ψ⟩
    typerho=str(rho.dtype)
    # If |ψ⟩ is complex then IF cond. is true
    if re.findall('^complex',typerho):
        # Calculating eigenvalues λ_i of |ψ⟩
        eigenvalues,eigenvectors,info=la.zheev(rho)
    # If |ψ⟩ is real then ELSE cond. is true
    else:
        eigenvalues,eigenvectors,info=la.dsyev(rho)
    # Calculating ln(λ_i)
    logeigenvalues=np.zeros([eigenvalues.shape[0]],dtype='float64')
    for i in range(0,eigenvalues.shape[0]):
        assert eigenvalues[i]>0.0,\
```

```
        "The eigenvalues of the matrix is coming less than equal to zero"
        # Calculating − ln(λᵢ)
        logeigenvalues[i]=(-1)*math.log(eigenvalues[i])
    return (eigenvalues,logeigenvalues)
```

Example

In this example, we are finding entanglement spectrum of a two qubit generalized Werner state $\rho(0.5)$.

```
from QuantumInformation import GatesTools as GT
gtobj=GT()
# Constructing the Werner state with mixing probability p = 0.5
state=gtobj.nWerner(0.5)
eigenvalues,logeigenvalues=entobj.entanglement_spectrum(state)
print(eigenvalues,logeigenvalues)
```

The preceding code prints the required eigenvalues and the entanglement spectrum as shown below.

```
[0.125 0.125 0.125 0.625] [2.07944154 2.07944154 2.07944154 0.47000363]
```

Similarly, we calculate the entanglement spectrum of the density matrix $\rho_1 \otimes \rho_2$, where density matrices ρ_1 and ρ_2 are already defined in Eq. [4.8] and Eq. [4.9], respectively.

```
# Constructing the density matrix ρ₁ ⊗ ρ₂
rho=np.zeros([4,4],dtype=np.complex_)
rho[0,0]=complex(0.16,0.0)
rho[0,1]=complex(-0.04,-0.16)
rho[0,2]=complex(-0.08,-0.04)
rho[0,3]=complex(-0.02,0.09)
rho[1,0]=complex(-0.04,0.16)
rho[1,1]=complex(0.24,0.0)
rho[1,2]=complex(0.06,-0.07)
rho[1,3]=complex(-0.12,-0.06)
rho[2,0]=complex(-0.08,0.04)
rho[2,1]=complex(0.06,0.07)
rho[2,2]=complex(0.24,0.0)
rho[2,3]=complex(-0.06,-0.24)
rho[3,0]=complex(-0.02,-0.09)
rho[3,1]=complex(-0.12,0.06)
rho[3,2]=complex(-0.06,0.24)
rho[3,3]=complex(0.36,0.0)
eigenvalues,logeigenvalues=entobj.entanglement_spectrum(rho)
print(eigenvalues,logeigenvalues)
```

The preceding code prints the required eigenvalues and the entanglement spectrum as shown below.

```
[0.01931653 0.0564194  0.2357345  0.68852957] [3.94679423 2.87494213
   1.44504911 0.37319701]
```

5.9   Residual entanglement for three qubits

Residual entanglement of a three qubit pure state is also known as three tangle [113]. We define the three tangle to be,

$$\tau_3 = \tau_{1(23)} - \tau_{12} - \tau_{13} \tag{5.25}$$

where τ can be defined as the square of concurrence. We can also define the tangle between one qubit (say the ith) and the rest of the qubits if the total state is pure. This is because of Schmidt decomposition. The tangle between qubit i and the rest is the one-tangle, given by $\tau_{i,(\text{rest of the spins})} = 4 \det(\rho_k)$, where ρ_k is the single qubit reduced density matrix got by partial tracing all other qubits except the kth qubit. This concept can be used to measure the entanglement in a three qubit system like the famous GHZ state or the W state. We can rewrite Eq. [5.25] as follows,

$$\tau_3 = 4 \det(\rho) - C_{12}^2 - C_{13}^2. \tag{5.26}$$

From the structure of the GHZ and W state, we see the contrasting behaviour of bipartite entanglement in them, this three tangle will be useful in differentiating these entanglement properties for such three qubit pure states.

5.9.1   For a real or complex state

`residual_entanglement_vec(state)`

Parameters

In/Out	Argument	Description
[in]	state	It is a real or complex array of dimension (0:7), which is the input pure state
[out]	res_tang	It returns the value of the residual entanglement

Implementation

```
def residual_entanglement_vec(self,state):
    """
    Calculation of residual entanglement for a three-qubit quantum state
    Parameters
        state : Real or complex 3-qubit state

    Returns
        res_tang : Residual entanglement value

    """
    # Checking whether entered state is a 3 qubit system
    assert state.shape[0]==8,"It is not a three qubit quantum system"
    # Calculating det(ρ)
```

```
det=np.linalg.det(self.partial_trace_vec(state,[1]))
# Calculating 4 det(ρ)
det=4*det
# Finally calculating, τ₃ = 4 det(ρ) − C²₁₂ − C²₁₃
res_tang=det-(self.concurrence_vec(state,1,2)**2)-\
(self.concurrence_vec(state,1,3)**2)
res_tang=abs(res_tang)
return res_tang
```

Example

In this example, we are finding the value of the residual entanglement of the three-qubit real state $|\psi_2\rangle$.

```
from QuantumInformation import QuantumMechanics as QM
qobj=QM()
# state2 will store |ψ₂⟩
state2=np.zeros([2**3],dtype='float64')
for i in range(0,state2.shape[0]):
    state2[i]=i+1
state2=qobj.normalization_vec(state2)
# Calculating τ₃
res_tang=entobj.residual_entanglement_vec(state2)
print(res_tang)
```

The preceding code prints the required value of the residual entanglement as,

```
1.8041124150158794e-16
```

Similarly, we calculate the value of the residual entanglement of the three-qubit complex state $|\psi_3\rangle$.

```
from QuantumInformation import QuantumMechanics as QM
qobj=QM()
# state2 will store |ψ₃⟩
state2=np.zeros([2**3],dtype=np.complex_)
for i in range(0,state2.shape[0]):
    state2[i]=complex(i,i+1)
state2=qobj.normalization_vec(state2)
# Calculating τ₃
res_tang=entobj.residual_entanglement_vec(state2)
print(res_tang)
```

The preceding code prints the required value of the residual entanglement as,

```
2.970954353025952e-17
```

5.9.2   For a real or complex density matrix

```
residual_entanglement_den(den)
```

Parameters

In/Out	Argument	Description
[in]	den	It is a real or complex array of dimension (0:7, 0:7), which is the input density matrix
[out]	res_tang	It returns the value of the residual entanglement

Implementation

```python
def residual_entanglement_den(self,den):
    """
    Calculation of residual entanglement for a three-qubit density matrix
    Parameters
        den : Real or complex 3-qubit density matrix

    Returns
        res_tang : Residual entanglement value
    """
    # Checking whether entered state is a 3-qubit system
    assert den.shape[0]==8,"It is not a three qubit quantum system"
    # Calculating det(ρ)
    det=np.linalg.det(self.partial_trace_den(den,[1]))
    # Calculating 4 det(ρ)
    det=4*det
    # Finally calculating, τ₃ = 4 det(ρ) − C²₁₂ − C²₁₃
    res_tang=det-(self.concurrence_den(den,1,2)**2)-\
    (self.concurrence_den(den,1,3)**2)
    res_tang=abs(res_tang)
    return res_tang
```

Example

In this example, we are finding the value of the residual entanglement of the three-qubit real density matrix $|\psi_2\rangle\langle\psi_2|$.

```python
from QuantumInformation import QuantumMechanics as QM
qobj=QM()
# state2 will store |ψ₂⟩⟨ψ₂|
state2=np.zeros([2**3],dtype='float64')
for i in range(0,state2.shape[0]):
    state2[i]=i+1
state2=qobj.normalization_vec(state2)
state2=qobj.outer_product_rvec(state2,state2)
# Calculating τ₃
res_tang=entobj.residual_entanglement_den(state2)
print(res_tang)
```

The preceding code prints the required value of the residual entanglement as,

```
1.8041124150158794e-16
```

Similarly, we calculate the value of the residual entanglement of the three-qubit complex density matrix $|\psi_3\rangle\langle\psi_3|$.

```
from QuantumInformation import QuantumMechanics as QM
qobj=QM()
# state2 will store |ψ₃⟩⟨ψ₃|
state2=np.zeros([2**3],dtype=np.complex_)
for i in range(0,state2.shape[0]):
    state2[i]=complex(i,i+1)
state2=qobj.normalization_vec(state2)
state2=qobj.outer_product_cvec(state2,state2)
# Calculating τ₃
res_tang=entobj.residual_entanglement_den(state2)
print(res_tang)
```

The preceding code prints the required value of the residual entanglement as,

```
2.970954353025952e-17
```

5.10 PPT criteria or the Peres positive partial transpose criteria

For any bipartite separable state ρ_{AB}, if we perform a partial transpose on one of the subsytems, say B, to obtain ρ^{T_B}, then we say ρ is separable if all the eigenvalues of ρ^{T_B} are non-negative. If ρ^{T_B} has even one negative eigenvalue, then ρ_{AB} is entangled. In the 2×2 and 2×3 dimensions the condition is both necessary and also sufficient [88, 94]. This method of entanglement detection is very important, as for a mixed state, the absence of Schmidt decomposition coupled with the non-uniqueness in the pure state decomposition of the mixed state makes PPT a noteworthy criteria. In the recipes below, we give the separability criteria for only 2×2 systems. We have given the example of PPT criteria using partial transpose routines as follows. In what follows, we will detect entanglement in the two-qubit generalized Werner state $\rho(0.5)$. As we already know that for $p > 1/3$ the Werner state is a non-separable state.

```
import scipy.linalg.lapack as la
from QuantumInformation import PartialTr as PT
from QuantumInformation import GatesTools as GT
import numpy as np

pobj=PT()
gtobj=GT()

# Constructing the Werner state ρ(0.5)
state=gtobj.nWerner(0.5)
sub_tr=[2]

# Partial transpose the subsystem 2
```

```
ptrans_state=pobj.ptranspose_den(state,sub_tr)

# Calculating the eigenvalues and eigenvectors of partial transposed state
eigenvalues, eigenvectors,info=la.dsyev(ptrans_state)

# flag=0 state not separable, and flag=1 state separable
flag=0
for i in range(0,eigenvalues.shape[0]):
    if eigenvalues[i] < 0.0:
        flag=1
if flag==1:
    print("Not separable")
elif flag==0:
    print("Separable")
```

The preceding code generates the following output,

```
Not separable
```

5.11   Reduction criteria

For any separable bipartite state ρ_{AB} such that $\rho_{AB} \in \mathcal{H}_{N_A \times N_B}$, where N_A and N_B are the dimensions of the subsystem spaces A and B, respectively. We then say that ρ_{AB} is separable only if the following operators are positive, meaning their eigenvalues are non-negative.

$$\rho^A \otimes \mathbb{I} - \rho_{AB} > 0, \tag{5.27}$$

$$\mathbb{I} \otimes \rho^B - \rho_{AB} > 0. \tag{5.28}$$

This criteria [114] is a set of two inequalities as shown above where ρ_A and ρ_B are the reduced density matrices of the subsystems A and B, respectively. In the recipes below, we give the separability criteria for only 2×2 systems. In this example, we will detect entanglement in two-qubit generalized Werner state $\rho(0.5)$.

```
from QuantumInformation import PartialTr as PT
from QuantumInformation import QuantumMechanics as QM
from QuantumInformation import GatesTools as GT
import scipy.linalg.lapack as la
import numpy as np

pobj=PT()
qobj=QM()
gtobj=GT()

# Constructing the Werner state ρ(0.5)
werner_state=gtobj.nWerner(0.5,2)
# Calculating reduced density matrix ρ^A
rdm=pobj.partial_trace_den(werner_state,[2])
# Constructing identity matrix 𝕀
identity=np.zeros([2,2],dtype='float64')
identity[0,0]=1.0
```

```
identity[1,1]=1.0
# Constructing ρ^A ⊗ 𝕀
mat=qobj.tensor_product_matrix(rdm,identity)
# Constructing ρ^A ⊗ 𝕀 − ρ(0.5)
final_mat=mat-werner_state
# Calculating eigenvalues of ρ^A ⊗ 𝕀 − ρ(0.5)
eigenvalues, eigenvectors, info = la.dsyev(final_mat)

# Checking the separability criteria of Werner state ρ(0.5)
status = 'Separable'

for i in range(0,len(eigenvalues)):
    if eigenvalues[i] < 0:
        status = 'Not separable'
        break
print(f"The state is {status}")
```

The preceding codes generates the following output

```
The state is Not separable
```

5.12   Complete example

In order to highlight some codes from these chapter, we first construct a six-qubit mixed state density matrix using Eq. [4.2] and Eq. [4.4] as shown below,

$$\rho_{123456} = \frac{3}{4}|\psi_2\rangle\langle\psi_2| + \frac{1}{4}|\psi_3\rangle\langle\psi_3|, \tag{5.29}$$

thereafter, we calculate block entropy value between blocks 145 and 236.

$$S(\rho_{123456}) = -Tr\left(\rho_{145}\log_2(\rho_{145})\right) = -Tr(\rho_{236}\log_2(\rho_{236})). \tag{5.30}$$

Next, we calculate the entanglement content between qubits 1 and 6, using the concurrence value C_{16}. Finally, we perform partial transpose operation on the subsystem B, which contains the qubits 2, 3 and 5 of the state ρ_{123456}, and determine whether the subsystem B is separable from subsystem A (A subsystem contains the qubits 1, 4 and 6) or not.

```
from QuantumInformation import PartialTr as PT
from QuantumInformation import GatesTools as GT
from QuantumInformation import Entanglement as ENT
from QuantumInformation import QuantumMechanics as QM
import scipy.linalg.lapack as la
import numpy as np

pobj=PT()
gtobj=GT()
entobj=ENT()
qobj=QM()
```

```python
# Constructing the state |ψ₂⟩⟨ψ₂|
state2=np.zeros([2**6],dtype='float64')
for i in range(0,state2.shape[0]):
    state2[i]=i+1
state2=qobj.normalization_vec(state2)
state2=qobj.outer_product_rvec(state2,state2)

# Constructing the state |ψ₃⟩⟨ψ₃|
state3=np.zeros([2**6],dtype=np.complex_)
for i in range(0,state3.shape[0]):
    state3[i]=complex(i,i+1)
state3=qobj.normalization_vec(state3)
state3=qobj.outer_product_cvec(state3,state3)

# Constructing the mixed state,  ρ₁₂₃₄₅₆ = ¾|ψ₂⟩⟨ψ₂| + ¼|ψ₃⟩⟨ψ₃|
mix_state=(0.75*state2)+(0.25*state3)

sub_tr=[2,3,6]

# Calculating block entropy,  S(ρ₁₂₃₄₅₆)
bentropy=entobj.block_entropy_den(mix_state,sub_tr)

print(f"The block entropy calculated out to be = {bentropy}")

# Calculating the concurrence  C₁₆
conc16=entobj.concurrence_den(state2,1,6)

print(f"The concurrence value calculated = {conc16}")

# Calculating the partial transposed state,  ρ₁₂₃₄₅₆^(T_B)
ptran_rho=pobj.ptranspose_den(mix_state,[2,3,5])

# Calculating the eigenvalues and eigenvectors of  ρ₁₂₃₄₅₆^(T_B)
eigenvalues, eigenvectors,info=la.zheev(ptran_rho)

# flag=0 state not separable, and flag=1 state separable
flag=0
for i in range(0,eigenvalues.shape[0]):
    if eigenvalues[i] < 0.0:
        flag=1
if flag==1:
    print("Not separable")
elif flag==0:
    print("Separable")
```

The preceding code generates the following output.

```
The block entropy calculated out to be = 0.08718991803231457
The concurrence value calculated = 0.011449016100176374
Not separable
```

6

One-Dimensional Quantum Spin-1/2 Chain Models

In this chapter we will give the numerical recipes for constructing the spin half Fermionic Hamiltonians in one dimension [115–118], basically a lattice of contiguous electrons along a line. Each electron at a given point on the one-dimensional lattice represents one qubit (here qubits are synonymous with spins) of information. Spin Hamiltonians or what is popularly called the spin chain models are prototypical in understanding the origin of magnetism and magnetic phases [119–121]. Not only that these spin chains at a temperature of zero Kelvin can be used to study and understand quantum phase transitions [80, 83], which unlike thermal phase transitions do not involve any thermal parameter. These Hamiltonians also prove as a fertile ground for testing various quantum information tasks not only theoretically but also experimentally [122–124]. We have given the numerical recipes for all possible interaction Hamiltonian's, and as we progress through the chapter, we will give some examples on how to construct some famous spin chains using our numerical recipes for the sake of illustration.

To understand further, we consider the simple but the well known Heisenberg spin chain model, a boon for people working in the area of magnetism. Using the Heisenberg model, the magnetic properties of many insulating crystals can be explained [125, 126]. The Hamiltonian of the Heisenberg model can be written as,

$$\hat{\mathcal{H}} = -J \sum_{i=1}^{N} \vec{\sigma}_i . \vec{\sigma}_{i+1}, \tag{6.1}$$

where J is the coupling constant and the periodic boundary condition is in effect, that is $\vec{\sigma}_{N+1} \equiv \vec{\sigma}_1$. Every spin is represented by a quantum operator acting upon the tensor product $(\mathbb{C}^2)^{\otimes N}$ as each site is associated with a complex two-dimensional Hilbert space. To make things more clear, let us take a 3 spin Heisenberg model, we can have two different types of boundary conditions, one is the periodic boundary condition (PBC) in which there is a wrapping up of the spins end-to-end and the other which is the open boundary condition (OBC) where there is no wrapping up and end spins are left free. Let us write the Hamiltonian for the model given in Eq. [6.1] as below, considering periodic boundary conditions with end to end wrap up of the form $\sigma_{N+1} \equiv \sigma_1$

$$\mathcal{H} = -J \sum_{i=1}^{3} \vec{\sigma}_i . \vec{\sigma}_{i+1} \tag{6.2}$$

$$= -J(\vec{\sigma}_1 . \vec{\sigma}_2 + \vec{\sigma}_2 . \vec{\sigma}_3 + \underbrace{\vec{\sigma}_3 . \vec{\sigma}_1}_{wrap\ up}), \tag{6.3}$$

$$= -J(\sigma_1^x \sigma_2^x + \sigma_1^y \sigma_2^y + \sigma_1^z \sigma_2^z + \sigma_2^x \sigma_3^x + \sigma_2^y \sigma_3^y + \sigma_2^z \sigma_3^z$$
$$\sigma_3^x \sigma_1^x + \sigma_3^y \sigma_1^y + \sigma_3^z \sigma_1^z) \tag{6.4}$$

Now for a model with open boundary conditions where there is no such condition of end to end wrap up like $\vec{\sigma}_{N+1} \equiv \vec{\sigma}_1$, we can write

$$\mathcal{H} = -J(\acute{\sigma}_1 . \acute{\sigma}_2 + \vec{\sigma}_2 . \vec{\sigma}_3) \tag{6.5}$$

$$= -J(\sigma_1^x \sigma_2^x + \sigma_1^y \sigma_2^y + \sigma_1^z \sigma_2^z + \sigma_2^x \sigma_3^x + \sigma_2^y \sigma_3^y + \sigma_2^z \sigma_3^z) \tag{6.6}$$

DOI: 10.1201/9781003285489-6

However, these are the PBCs and OBCs for nearest-neighbour interaction, if we have r nearest neighbours interacting, there will be r such end to end wrap up terms of the form $\sigma_{N+r} \equiv \sigma_r$ for $r = 1, 2, \ldots$. It is also important to note that $\sigma_i^z = \mathbb{I}^{\otimes i-1} \otimes \sigma^z \otimes \mathbb{I}^{\otimes N-i}$, where $\mathbb{I}$ is the 2×2 identity matrix. This is so because, the total Hamiltonian for N spins will be a $2^N \times 2^N$ matrix in the $(\mathbb{C}^2)^{\otimes N}$ – dimensional Hilbert space. All this implies that the Pauli operator z with a subscript i which is σ_i^z will only act on the qubit at site i. At each site i, there is a quantum mechanical spin half particle either pointing along "up ($|\uparrow\rangle$)" or "down ($|\downarrow\rangle$)" which are respectively denoted by $|0\rangle$ and $|1\rangle$. Having the above discussed facts in mind, we can proceed further. All the recipes given below are for both PBC's and OBC's which can be chosen as a user defined variable while execution. The user also can choose till how many neighbours he or she needs to build the spin chain. Also note that in this chapter we are using σ matrices instead of the naive spin matrix S, bearing in mind the relationship between them as $S_i = \frac{1}{2}\sigma_i$ (with $\hbar = 1$), with $i = x, y, z$.

Note that for all Hamiltonians from Section (6.1) to Section (6.3), the output of an example contains only the integer part of the matrix elements; this is done to accommodate the output in the given space. However, for the Section (6.4) which is the DM interaction, only integer part of the imaginary elements are displayed for the same reason. In this chapter, all examples are given for three spins. In the codes given in this chapter we have represented the interaction parameters like magnetic field, interaction strength (direct and anti symmetric) by an array of dimension equal to number of spins in the system.

In the codes given for the interaction of the spins with magnetic field, we have considered B to be an array of dimension equal to the number of spins in the system. This also gives us the freedom to choose an inhomogenenous magnetic field at different sites, meaning, the ith entry of this array will be the magnitude of the applied magnetic field at the ith site. If the magnetic field is homogeneous, all entries of the array B will be same. Similarly, for spin-spin interactions, we have chosen the interaction parameter to be an array of dimension equal to the number of spins in the system. The ith entry of this array will be the interaction strength between the ith and $(i+r)^{th}$ spin. If the interaction is homogeneous, all the entries of the corresponding interaction parameter will be same. However, if random interactions are present between the spins, the array will be filled with different numbers (drawn from any probability distribution) corresponding to each pair of spins. You can understand more about the usage as you see the examples given as we evolve through the chapter.

To import the class **QuantumMechanics** from the **QuantumInformation** library, and creating the object of the class in your Python code can be done as follows,

```
# Importing the class Hamiltonian from the QuantumInformation library
from QuantumInformation import Hamiltonian as ham

# creating the object of the class
hamobj=ham()
```

6.1   Hamiltonian of spins interacting with an external magnetic field

The interaction Hamiltonian of a linear array of spin half electrons interacting with an external magnetic field of strength B in general is given by,

$$H = \sum_{i=1}^{N} B_i \sigma_i^j, \tag{6.7}$$

where N is the total number of qubits present in the system and σ^j are the Pauli spin matrices, depending on which direction, the magnetic field is oriented, j can be accordingly labeled as x, y, z as follows. Note that B_i is inside summation to accommodate a general inhomogeneous magnetic field.

6.1.1 Hamiltonian of spins interacting with an external magnetic field in the X direction

In this case we choose $j = x$ in Eq. [6.7] and thereby, the interaction Hamiltonian of a linear array of spins interacting with the external magnetic field in the X direction is given by,

$$H = \sum_{i=1}^{N} B_i \sigma_i^x, \tag{6.8}$$

field_xham(N,mode='homogeneous',B=1)

Parameters

In/Out	Argument	Description
[in]	N	It is the total number of spins
[in]	mode	It specifies whether the external magnetic field is homogeneous or inhomogeneous If mode = 'homogeneous', then we have homogeneous magnetic ($B_i = B$). It is the default value If mode = 'inhomogeneous', then we have inhomogeneous magnetic field
[in]	B	If the mode = 'homogeneous', B takes in a constant value, and the default value is 1 If the mode = 'inhomogeneous', we pass a list holding the B_i value to parameter B. The ith entry of the list is the strength of the applied magnetic field along X direction on the ith spin, therefore, the list will contain N such entries
[out]	xham	It is an array of dimension (0:2**N−1,0:2**N−1), which is the required Hamiltonian

Implementation

```
def field_xham(self,N,mode='homogeneous',B=1):
    """
    Constructs Hamiltonian of external magnetic field in X direction
    Input
        N: number of spins
        mode: 'homogeneous' or 'inhomogeneous' magnetic field
        B: it list of value if mode='inhomogeneous', and constant if
            mode='homogeneous'
    Output
```

```
        xham: Hamiltonian
    """
    # Only two modes are allowed, homogeneous  and inhomogeneous
    assert mode == 'homogeneous' or mode == 'inhomogeneous',\
    "Entered mode is invalid"
    # Number of spins N ≥ 1
    assert N >= 1, "number of spins entered is not correct"
    # Entering the homogeneous mode
    if mode == 'homogeneous':
        # It will store ∑_{i=1}^{N} Bσ_i^x
        xham=np.zeros([2**N,2**N],dtype='float64')
        # Row index of ∑_{i=1}^{N} Bσ_i^x
        for i in range(0,2**N,1):
            # Column index of ∑_{i=1}^{N} Bσ_i^x
            for j in range(0,2**N,1):
                # Decimal to binary decision
                bvec=qobj.decimal_binary(j,N)
                sum1=0.0
                for k in range(0,N):
                    bb=np.copy(bvec)
                    # Matrix operation, σ^z|a⟩ = |1 − a⟩, where a = {0,1}
                    bb[k]=1-bb[k]
                    row=qobj.binary_decimal(bb)
                    if row == i:
                        sum1=sum1+B
                xham[i,j]=sum1
    else:
        # Checking whether there are N values of B_i
        assert len(B)==N,\
        "The entered values of magnetic strengths are not equal to number
of spins"
            # It will store ∑_{i=1}^{N} B_iσ_i^x
        xham=np.zeros([2**N,2**N],dtype='float64')
        for i in range(0,2**N,1):
            # Row index of ∑_{i=1}^{N} B_iσ_i^x
            for j in range(0,2**N,1):
                # Column index of ∑_{i=1}^{N} B_iσ_i^x
                bvec=qobj.decimal_binary(j,N)
                sum1=0.0
                for k in range(0,N):
                    bb=np.copy(bvec)
                    # Matrix operation, σ^z|a⟩ = |1 − a⟩, where a = {0,1}
                    bb[k]=1-bb[k]
                    row=qobj.binary_decimal(bb)
                    if row == i:
                        sum1=sum1+B[k]
                xham[i,j]=sum1
    return xham
```

Example

Here, we are finding the Hamiltonian of spins interacting with homogeneous magnetic field in X direction.

```
hamx=hamobj.field_xham(3)
print(hamx)
```

The preceding code prints the matrix elements of the required Hamiltonian.

```
[[0. 1. 1. 0. 1. 0. 0. 0.]
 [1. 0. 0. 1. 0. 1. 0. 0.]
 [1. 0. 0. 1. 0. 0. 1. 0.]
 [0. 1. 1. 0. 0. 0. 0. 1.]
 [1. 0. 0. 0. 0. 1. 1. 0.]
 [0. 1. 0. 0. 1. 0. 0. 1.]
 [0. 0. 1. 0. 1. 0. 0. 1.]
 [0. 0. 0. 1. 0. 1. 1. 0.]]
```

6.1.2 Hamiltonian of spins interacting with an external magnetic field in the Y direction

In this case we choose $j = y$ in Eq. [6.7] and thereby, the interaction Hamiltonian of a linear array of spins interacting with the external magnetic field in the Y direction is given by,

$$H = \sum_{i=1}^{N} B_i \sigma_i^y. \tag{6.9}$$

```
field_yham(N,mode='homogeneous',B=1)
```

Parameters

In/Out	Argument	Description
[in]	N	It is the total number of spins
[in]	mode	It specifies whether the external magnetic field is homogeneous or inhomogeneous If mode = 'homogeneous', we have homogeneous magnetic $(B_i = B)$. It is the default value If mode = 'inhomogeneous', we have inhomogeneous magnetic field
[in]	B	If the mode = 'homogeneous', B takes in a constant value, and the default value is 1 If the mode = 'inhomogeneous', we pass a list holding the B_i value to parameter B. The ith entry of the list is the strength of the applied magnetic field along X direction on the ith spin, therefore, the list will contain N such entries
[out]	yham	It is an array of dimension (0:2**N−1,0:2**N−1), which is the required Hamiltonian

Implementation

```python
def field_yham(self,N,mode='homogeneous',B=1):
    """

    Constructs Hamiltonian of external magnetic field in Y direction
    Input
        N: number of spins
        mode: 'homogeneous' or 'inhomogeneous' magnetic field
        B: it list of value if mode='inhomogeneous', and constant if
            mode='homogeneous'
    Output
        yham: Hamiltonian
    """

    # Only two modes are allowed, homogeneous  and inhomogeneous
    assert mode == 'homogeneous' or mode == 'inhomogeneous',\
    "Entered mode is invalid"
    # Number of spins N >= 1
    assert N >= 1, "number of spins entered is not correct"
    if mode == 'homogeneous':
        # It will store, $\sum_{i=1}^{N} B\sigma_i^y$
        yham=np.zeros([2**N,2**N],dtype=np.complex_)
        # Row index of $\sum_{i=1}^{N} B\sigma_i^y$
        for i in range(0,2**N,1):
            # Column index of $\sum_{i=1}^{N} B\sigma_i^y$
            for j in range(0,2**N,1):
                # Decimal to binary conversions
                bvec=qobj.decimal_binary(j,N)
                sum1=0.0
                for k in range(0,N):
                    bb=np.copy(bvec)
                    # Matrix operation, $\sigma^y|a\rangle = (-i)^a|1-a\rangle$, where $a = \{0,1\}$
                    bb[k]=1-bb[k]
                    row=qobj.binary_decimal(bb)
                    if row == i:
                        sum1=sum1+(B*complex(0,1)*((-1)**bvec[k]))
                yham[i,j]=sum1
    else:
        # Checking whether there are $N$ values of $B_i$
        assert len(B)==N,\
        "The entered values of magnetic strengths are not equal to number
    of spins"
        # It will store, $\sum_{i=1}^{N} B_i\sigma_i^y$
        yham=np.zeros([2**N,2**N],dtype=np.complex_)
        for i in range(0,2**N,1):
            # Row index, $\sum_{i=1}^{N} B_i\sigma_i^y$
            for j in range(0,2**N,1):
                # Column index, $\sum_{i=1}^{N} B_i\sigma_i^y$
                bvec=qobj.decimal_binary(j,N)
                sum1=0.0
                for k in range(0,N):
```

```
            bb=np.copy(bvec)
            # Matrix operation, σʸ|a⟩ = (−i)ᵃ|1 − a⟩, where a = {0,1}
            bb[k]=1-bb[k]
            row=qobj.binary_decimal(bb)
            if row == i:
                sum1=sum1+(B[k]*complex(0,1)*((-1)**bvec[k]))
        yham[i,j]=sum1
    return yham
```

Example

Here, we are finding the Hamiltonian of spins interacting with homogeneous magnetic field in Y direction.

```
hamy=hamobj.field_yham(3)
print(hamy)
```

The preceding code prints the matrix elements of the required Hamiltonian.

```
[[0.+0.j 0.-1.j 0.-1.j 0.+0.j 0.-1.j 0.+0.j 0.+0.j 0.+0.j]
 [0.+1.j 0.+0.j 0.+0.j 0.-1.j 0.+0.j 0.-1.j 0.+0.j 0.+0.j]
 [0.+1.j 0.+0.j 0.+0.j 0.-1.j 0.+0.j 0.+0.j 0.-1.j 0.+0.j]
 [0.+0.j 0.+1.j 0.+1.j 0.+0.j 0.+0.j 0.+0.j 0.+0.j 0.-1.j]
 [0.+1.j 0.+0.j 0.+0.j 0.+0.j 0.+0.j 0.-1.j 0.-1.j 0.+0.j]
 [0.+0.j 0.+1.j 0.+0.j 0.+0.j 0.+1.j 0.+0.j 0.+0.j 0.-1.j]
 [0.+0.j 0.+0.j 0.+1.j 0.+0.j 0.+1.j 0.+0.j 0.+0.j 0.-1.j]
 [0.+0.j 0.+0.j 0.+0.j 0.+1.j 0.+0.j 0.+1.j 0.+1.j 0.+0.j]]
```

6.1.3 Hamiltonian of spins interacting with an external magnetic field in the Z direction

In this case we choose $j = z$ in Eq. [6.7] and thereby, the interaction Hamiltonian of a linear array of spins interacting with the external magnetic field in the Z direction is given by,

$$H = \sum_{i=1}^{N} B_i \sigma_i^z. \tag{6.10}$$

```
field_zham(N,mode='homogeneous',B=1)
```

Parameters

In/Out	Argument	Description
[in]	N	It is the total number of spins
[in]	mode	It specifies whether the external magnetic field is homogeneous or inhomogeneous If mode = 'homogeneous', we have homogeneous magnetic ($B_i = B$). It is the default value If mode = 'inhomogeneous', we have inhomogeneous magnetic field
[in]	B	If the mode = 'homogeneous', B takes in a constant value, and the default value is 1 If the mode = 'inhomogeneous', we pass a list holding the B_i value to parameter B. The ith entry of the list is the strength of the applied magnetic field along X direction on the ith spin, therefore, the list will contain N such entries
[out]	zham	It is an array of dimension $(0{:}2{**}N{-}1, 0{:}2{**}N{-}1)$, which is the required Hamiltonian

Implementation

```
def field_zham(self,N,mode='homogeneous',B=1):
    """
    Constructs Hamiltonian of external magentic field in Z direction
    Input
        N: number of spins
        mode: 'homogeneous' or 'inhomogeneous' magnetic field
        B: it list of value if mode='inhomogeneous', and constant if
           mode='homogeneous'
    Output
        zham: Hamiltonian
    """

    # Only two modes are allowed, homogeneous  and inhomogeneous
    assert mode == 'homogeneous' or mode == 'inhomogeneous',\
    "Entered mode is invalid"
    # Number of spins N >= 1
    assert N >= 1, "number of spins entered is not correct"
    zham=np.zeros([2**N,2**N],dtype='float64')
    if mode == 'homogeneous':
        # Constructing Hamiltonian, $\sum_{i=1}^{N} B\sigma_i^z$
        for i in range(0,2**N):
            sum1=0.0
            bvec=qobj.decimal_binary(i,N)
            for k in range(0,N):
                sum1=sum1+(((-1)**bvec[k])*B)
            zham[i,i]=sum1
    else:
```

```
# Checking whether there are N values of B_i
assert len(B)==N,\
"The entered values of magnetic strengths are not equal to number
of spins"
    # Constructing Hamiltonian, \sum_{i=1}^{N} B_i \sigma_i^z
    for i in range(0,2**N):
        sum1=0.0
        bvec=qobj.decimal_binary(i,N)
        for k in range(0,N):
            sum1=sum1+(((-1)**bvec[k])*B[k])
        zham[i,i]=sum1
return zham
```

Example

Here, we are finding the Hamiltonian of spins interacting with inhomogeneous magnetic field in Z direction.

```
hamz=hamobj.field_zham(3)
print(hamz)
```

The preceding code prints the matrix elements of the required Hamiltonian.

```
[[ 3.  0.  0.  0.  0.  0.  0.  0.]
 [ 0.  1.  0.  0.  0.  0.  0.  0.]
 [ 0.  0.  1.  0.  0.  0.  0.  0.]
 [ 0.  0.  0. -1.  0.  0.  0.  0.]
 [ 0.  0.  0.  0.  1.  0.  0.  0.]
 [ 0.  0.  0.  0.  0. -1.  0.  0.]
 [ 0.  0.  0.  0.  0.  0. -1.  0.]
 [ 0.  0.  0.  0.  0.  0.  0. -3.]]
```

6.2   Hamiltonian for the direct exchange spin–spin interaction

The Hamiltonian for the quantum mechanical spin-spin exchange interaction is given by:

$$H = \sum_{i=1}^{N} J_r^i \sigma_i^j \sigma_{i+r}^j. \tag{6.11}$$

Here J_r^i is the interaction strength (which can be both positive and negative) which is also called the coupling constant between the ith and the $(i+r)^{th}$ spins with r representing the rth neighbouring interaction. Here N is the total number of spins (qubits) present in the system and σ^j are the Pauli spin matrices, where $j = x, y, z$. The notation σ_{i+r}^j denotes the Pauli spin matrices corresponding to the rth neighbouring spin interacting with the spin represented by σ_i^j at the ith location. Note that J_r^i may be a constant or it can be chosen from any probability distribution to induce bond randomness. In what follows below, we discuss three cases.

6.2.1   Spin–spin interaction in X direction

The Hamiltonian for the spin–spin interaction in the X direction is obtained by, choosing $j = x$ in Eq. [6.11] as

$$H = \sum_{i=1}^{N} J_r^i \sigma_i^x \sigma_{i+r}^x. \tag{6.12}$$

```
ham_xx(N,nn=1,mode='homogeneous',jx=1,condition='periodic')
```

Parameters

In/Out	Argument	Description
[in]	N	It is the total number of spins
[in]	nn	nn stores the value of r describing the rth neighbour interaction of the Hamiltonian. The default value is equal to 1 (nearest neighbour interaction)
[in]	mode	It specifies whether the interaction strength is uniform or non-uniform If mode = 'homogeneous', we have an uniform interaction ($J_r^i = J$). It is the default value If mode = 'inhomogeneous', then we have non-uniform interaction
[in]	jx	If the mode = 'homogeneous', then jx takes in a constant value, and the default value is 1. If the mode = 'inhomogeneous', then we pass a list holding the J_r^i value to parameter jx. The ith entry represents the interaction strength between the ith and $(i + r)^{th}$ spin
[in]	condition	It defines the type of boundary condition if condition = 'periodic', periodic boundary conditions will be considered. It is the default value if condition = 'open', open boundary conditions will be considered
[out]	ham	It is an array of dimension (0:2**N−1,0:2**N−1), which is the required Hamiltonian

Implementation

```
def ham_xx(self,N,nn=1,mode='homogeneous',jx=1,condition='periodic'):
    """
    Constructs Hamiltonian of spin-spin interaction in X direction
    Input
        N: number of spins
        nn: specifying rth interaction
        mode: 'homogeneous' or 'inhomogeneous' magnetic field
        jx: it list of values if mode='inhomogeneous', and constant if
```

```
        mode='homogeneous'
    condition:  defining boundary conditions of the Hamiltonian
Output
    ham: Hamiltonian
"""
# Checking entered boundary condition, PBC and OBC
assert condition == 'periodic' or condition == 'open'
# Number of interaction depending on the boundary condition
if condition == 'periodic':
    kn=N
else:
    kn=N-nn
# Only two modes are allowed, homogeneous and inhomogeneous
assert mode == 'homogeneous' or mode == 'inhomogeneous',\
"Entered mode is invalid"
# Checking number of spins $N \geq 1$
assert N >= 1, "number of spins entered is not correct"
# Checking valid $r^{th}$ interaction
assert nn<=N-1 and nn>=1,"Not valid interaction"
if mode == 'homogeneous':
    ham=np.zeros([2**N,2**N],dtype='float64')
    col2=np.zeros([N,1])
    # Row index of matrix, $\sum_{i=1}^{N} J_r \sigma_i^x \sigma_{i+r}^x$
    for i in range(0,2**N):
        # Column index of matrix, $\sum_{i=1}^{N} J_r \sigma_i^x \sigma_{i+r}^x$
        for j in range(0,2**N):
            col=qobj.decimal_binary(j,N)
            for k in range(0,kn):
                k1=k+nn
                # For wrapping the Hamiltonian
                if k1>=N:
                    k1=k1-N
                col2=col.copy()
                col2[k]=1-col2[k]
                col2[k1]=1-col2[k1]
                dec=qobj.binary_decimal(col2)
                if dec==i:
                    inn=1
                else:
                    inn=0
                ham[i,j]=ham[i,j]+(inn*jx)
else:
    # Checking number of $J_r^i$ interactions are valid according to BC
    if condition=='periodic':
        assert len(jx)==N,\
        "The entered values of magnetic strengths are not equal to
number of spins"
    else:
        assert len(jx)==N-nn,\
```

```
          "The entered values of magnetic strengths are not equal to
     number of spins"
          ham=np.zeros([2**N,2**N],dtype='float64')
          col2=np.zeros([N,1])
          # Row index of matrix, ∑_{i=1}^{N} J_r^i σ_i^x σ_{i+r}^x
          for i in range(0,2**N):
              # Column index of matrix, ∑_{i=1}^{N} J_r^i σ_i^x σ_{i+r}^x
              for j in range(0,2**N):
                  col=qobj.decimal_binary(j,N)
                  for k in range(0,kn):
                      k1=k+nn
                      # For wrapping the Hamiltonian
                      if k1>=N:
                          k1=k1-N
                      col2=col.copy()
                      col2[k]=1-col2[k]
                      col2[k1]=1-col2[k1]
                      dec=qobj.binary_decimal(col2)
                      if dec==i:
                          inn=1
                      else:
                          inn=0
                      ham[i,j]=ham[i,j]+(inn*jx[k])
     return ham
```

Example

In this example, we are finding the Hamiltonian of spin–spin homogeneous interaction $(J_r=(1,1,1))$ with periodic boundary conditions in X direction.

```
ham=hamobj.ham_xx(3)
print(ham)
```

The preceding code prints the matrix elements of the required Hamiltonian.

```
[[0. 0. 0. 1. 0. 1. 1. 0.]
 [0. 0. 1. 0. 1. 0. 0. 1.]
 [0. 1. 0. 0. 1. 0. 0. 1.]
 [1. 0. 0. 0. 0. 1. 1. 0.]
 [0. 1. 1. 0. 0. 0. 0. 1.]
 [1. 0. 0. 1. 0. 0. 1. 0.]
 [1. 0. 0. 1. 0. 1. 0. 0.]
 [0. 1. 1. 0. 1. 0. 0. 0.]]
```

In the following example, we are finding the Hamiltonian of spin–spin inhomogeneous interaction ($J_r^1 = 0.1$, and $J_r^2 = 0.3$) with open boundary conditions in X direction.

```
ham=hamobj.ham_xx(3,mode='inhomogeneous',jx=[0.1,0.3],condition='open')
print(ham)
```

The preceding codes prints the required Hamiltonian as shown below.

```
[[0.   0.   0.   0.3 0.   0.   0.1 0. ]
 [0.   0.   0.3 0.   0.   0.   0.   0.1]
 [0.   0.3 0.   0.   0.1 0.   0.   0. ]
 [0.3 0.   0.   0.   0.   0.1 0.   0. ]
 [0.   0.   0.1 0.   0.   0.   0.   0.3]
 [0.   0.   0.   0.1 0.   0.   0.3 0. ]
 [0.1 0.   0.   0.   0.   0.3 0.   0. ]
 [0.   0.1 0.   0.   0.3 0.   0.   0. ]]
```

6.2.2 Spin–spin interaction in Y direction

The Hamiltonian for the spin–spin exchange interaction in the Y direction is obtained by, choosing $j = y$ in Eq. [6.11] as

$$H = \sum_{i=1}^{N} J_r^i \sigma_i^y \sigma_{i+r}^y. \tag{6.13}$$

```
ham_yy(N,nn=1,mode='homogeneous',jy=1,condition='periodic')
```

Parameters

In/Out	Argument	Description
[in]	N	It is the total number of spins
[in]	nn	nn stores the value of r describing the rth neighbour interaction of the Hamiltonian. The default value is equal to 1 (nearest neighbour interaction)

In/Out	Argument	Description
[in]	mode	It specifies whether the interaction strength is uniform or non-uniform If mode = 'homogeneous', then we have an uniform interaction ($J_r^i = J$). It is the default value If mode = 'inhomogeneous', then we have non-uniform interaction
[in]	jy	If the mode = 'homogeneous', jx takes in a constant value, and the default value is 1. If the mode = 'inhomogeneous', then we pass a list holding the J_r^i value to parameter jx. The ith entry represents the interaction strength between the ith and $(i+r)^{th}$ spin
[in]	condition	It defines the type of boundary condition if condition = 'periodic', periodic boundary conditions will be considered. It is the default value if condition = 'open', open boundary conditions will be considered
[out]	ham	It is an array of dimension (0:2**N−1,0:2**N−1), which is the required Hamiltonian

Implementation

```
def ham_yy(self,N,nn=1,mode='homogeneous',jy=1,condition='periodic'):
    """

    Constructs Hamiltonian of spin-spin interaction in Y direction
    Input
        N: number of spins
        nn: specifying rth interaction
        mode: 'homogeneous' or 'inhomogeneous' magnetic field
        jy: it list of values if mode='inhomogeneous', and constant if
            mode='homogeneous'
        condition:  defining boundary conditions of the Hamiltonian
    Output
        ham: Hamiltonian

    """

    # Checking entered boundary condition, PBC and OBC
    assert condition == 'periodic' or condition == 'open'
    # Number of interaction depending on the boundary condition
    if condition == 'periodic':
        kn=N
    else:
        kn=N-nn
    # Only two modes are allowed, homogeneous and inhomogeneous
    assert mode == 'homogeneous' or mode == 'inhomogeneous',\
    "Entered mode is invalid"
```

```
# Checking number of spins N ≥ 1
assert N >= 1, "number of spins entered is not correct"
# Checking valid rᵗʰ interaction
assert nn<=N-1 and nn>=1,"Not valid interaction"
ham=np.zeros([2**N,2**N],dtype='float64')
col2=np.zeros([N,1])
if mode == 'homogeneous':
    # Row index of matrix, ∑_{i=1}^N J_r σ_i^y σ_{i+r}^y
    for i in range(0,2**N):
        # Column index of matrix, ∑_{i=1}^N J_r σ_i^y σ_{i+r}^y
        for j in range(0,2**N):
            col=qobj.decimal_binary(j,N)
            for k in range(0,kn):
                k1=k+nn
                # For wrapping the Hamiltonian
                if k1>=N:
                    k1=k1-N
                col2=col.copy()
                col2[k]=1-col2[k]
                col2[k1]=1-col2[k1]
                if col2[k]==col2[k1]:
                    ind=-1
                else:
                    ind=1
                dec=qobj.binary_decimal(col2)
                if dec==i:
                    inn=1
                else:
                    inn=0
                ham[i,j]=ham[i,j]+(inn*jy*ind)
else:
    # Checking number of J_r^i interactions are valid according to BC
    if condition=='periodic':
        assert len(jy)==N,\
        "The entered values of magnetic strengths are not equal to
number of spins"
    else:
        assert len(jy)==N-nn,\
        "The entered values of magnetic strengths are not equal to
number of spins"
    # Row index of matrix, ∑_{i=1}^N J_r^i σ_i^y σ_{i+r}^y
    for i in range(0,2**N):
        # Column index of matrix, ∑_{i=1}^N J_r^i σ_i^y σ_{i+r}^y
        for j in range(0,2**N):
            col=qobj.decimal_binary(j,N)
            for k in range(0,kn):
                k1=k+nn
                # For wrapping the Hamiltonian
                if k1>=N:
                    k1=k1-N
```

```
col2=col.copy()
col2[k]=1-col2[k]
col2[k1]=1-col2[k1]
if col2[k]==col2[k1]:
    ind=-1
else:
    ind=1
dec=qobj.binary_decimal(col2)
if dec==i:
    inn=1
else:
    inn=0
ham[i,j]=ham[i,j]+(inn*jy[k]*ind)
return ham
```

Example

In this example, we are finding the Hamiltonian of three qubit spin–spin interaction in the Y direction with open boundary conditions and random interactions. The Hamiltonian has nearest-neighbour interaction, and the strength of the random interactions are $J_1^1 = 0.1$ and $J_1^2 = -0.3$.

```
ham_yy=hamobj.ham_yy(3,nn=1,mode='inhomogeneous',\
                jy=[0.1,-0.3],condition='open')
print(ham_yy)
```

The preceding code prints the required Hamiltonian matrix.

```
[[ 0.    0.    0.    0.3   0.    0.   -0.1   0. ]
 [ 0.    0.   -0.3   0.    0.    0.    0.   -0.1]
 [ 0.   -0.3   0.    0.    0.1   0.    0.    0. ]
 [ 0.3   0.    0.    0.    0.    0.1   0.    0. ]
 [ 0.    0.    0.1   0.    0.    0.    0.    0.3]
 [ 0.    0.    0.    0.1   0.    0.   -0.3   0. ]
 [-0.1   0.    0.    0.    0.   -0.3   0.    0. ]
 [ 0.   -0.1   0.    0.    0.3   0.    0.    0. ]]
```

6.2.3   Spin–spin interaction in Z direction

The Hamiltonian for the spin-spin exchange interaction in the Z direction is obtained by, choosing $j = z$ in Eq. [6.11] as

$$H = \sum_{i=1}^{N} J_r^i \sigma_i^z \sigma_{i+r}^z. \tag{6.14}$$

```
ham_zz(N,nn=1,mode='homogeneous',jz=1,condition='periodic')
```

Parameters

In/Out	Argument	Description
[in]	N	It is the total number of spins
[in]	nn	nn stores the value of r describing the rth neighbour interaction of the Hamiltonian. The default value is equal to 1 (nearest neighbour interaction)
[in]	mode	It specifies whether the interaction strength is uniform or non-uniform If mode = 'homogeneous', we have an uniform interaction $(J_r^i = J)$. It is the default value If mode = 'inhomogeneous', we have non-uniform interaction
[in]	jz	If the mode = 'homogeneous', jz takes in a constant value and the default value is 1. If the mode = 'inhomogeneous', then we pass a list holding the J_r^i value to parameter jz. The ith entry represents the interaction strength between the ith and $(i + r)^{th}$ spin
[in]	condition	It defines the type of boundary condition if condition = 'periodic', periodic boundary conditions will be considered. It is the default value. if condition = 'open', open boundary conditions will be considered
[out]	ham	It is an array of dimension (0:2**N−1,0:2**N−1), which is the required Hamiltonian

Implementation

```
def ham_zz(self,N,nn=1,mode='homogeneous',jz=1,condition='periodic'):
    """
    Constructs Hamiltonian of spin-spin interaction in Z direction
    Input
        N: number of spins
        nn: specifying rth interaction
        mode: 'homogeneous' or 'inhomogeneous' magnetic field
        jz: it list of values if mode='inhomogeneous', and constant if
            mode='homogeneous'
        condition:  defining boundary conditions of the Hamiltonian
    Output
        ham: Hamiltonian
    """
    # Checking entered boundary condition, PBC and OBC
    assert condition == 'periodic' or condition == 'open'
    # Number of interaction depending on the boundary condition
    if condition == 'periodic':
        kn=N
```

```
    else:
        kn=N-nn
    # Only two modes are allowed, homogeneous and inhomogeneous
    assert mode == 'homogeneous' or mode == 'inhomogeneous',\
    "Entered mode is invalid"
    # Checking number of spins N ≥ 1
    assert N >= 1, "number of spins entered is not correct"
    # Checking valid r^{th} interaction
    assert nn<=N-1 and nn>=1,"Not valid interaction"
    ham=np.zeros([2**N,2**N],dtype='float64')
    if mode == 'homogeneous':
        # Row index of matrix, ∑_{i=1}^{N} J_r σ_i^z σ_{i+r}^z
        for i in range(0,2**N):
            # Column index of matrix, ∑_{i=1}^{N} J_r σ_i^z σ_{i+r}^z
            for j in range(0,2**N):
                col=qobj.decimal_binary(j,N)
                for k in range(0,kn):
                    k1=k+nn
                    # For wrapping the Hamiltonian
                    if k1>=N:
                        k1=k1-N
                    egv1=1-2*col[k]
                    egv1=int(egv1)
                    egv2=1-2*col[k1]
                    egv2=int(egv2)
                    if i==j:
                        inn=1
                    else:
                        inn=0
                    ham[i,j]=ham[i,j]+(inn*jz*egv1*egv2)
    else:
        # Checking number of J_r^i interactions are valid according to BC
        if condition=='periodic':
            assert len(jz)==N,\
            "The entered values of magnetic strengths are not equal to
number of spins"
        else:
            assert len(jz)==N-nn,\
            "The entered values of magnetic strengths are not equal to
number of spins"
        # Row index of matrix, ∑_{i=1}^{N} J_r^i σ_i^z σ_{i+r}^z
        for i in range(0,2**N):
            # Column index of matrix, ∑_{i=1}^{N} J_r^i σ_i^z σ_{i+r}^z
            for j in range(0,2**N):
                col=qobj.decimal_binary(j,N)
                for k in range(0,kn):
                    k1=k+nn
                    # For wrapping the Hamiltonian
                    if k1>=N:
                        k1=k1-N
```

```
                    egv1=1-2*col[k]
                    egv1=int(egv1)
                    egv2=1-2*col[k1]
                    egv2=int(egv2)
                    if i==j:
                        inn=1
                    else:
                        inn=0
                    ham[i,j]=ham[i,j]+(inn*jz[k]*egv1*egv2)
        return ham
```

Example

In this example, we are finding the Hamiltonian of the three qubit spin-spin homogeneous interaction ($J_r^i = 2$, for all $i = \{1, 2, 3\}$) with periodic boundary conditions in Z direction. The Hamiltonian has nearest-neighbour interactions ($r = 1$).

```
ham_zz=hamobj.ham_zz(3,jz=2)
print(ham_zz)
```

The preceding code prints the required Hamiltonian as shown below.

```
[[ 6.  0.  0.  0.  0.  0.  0.  0.]
 [ 0. -2.  0.  0.  0.  0.  0.  0.]
 [ 0.  0. -2.  0.  0.  0.  0.  0.]
 [ 0.  0.  0. -2.  0.  0.  0.  0.]
 [ 0.  0.  0.  0. -2.  0.  0.  0.]
 [ 0.  0.  0.  0.  0. -2.  0.  0.]
 [ 0.  0.  0.  0.  0.  0. -2.  0.]
 [ 0.  0.  0.  0.  0.  0.  0.  6.]]
```

6.2.4 Variants of the Ising model

Now, we will see several variants of the Ising model which is being used in condensed matter physics.

6.2.4.1 The homogeneous Ising model in a non-zero uniform magnetic field

The Hamiltonian of Ising model in a non-zero uniform magnetic field is given by,

$$H = -J \sum_{i=1}^{N} \sigma_i^z \sigma_{i+1}^z - B \sum_{i=1}^{N} \sigma_i^z. \tag{6.15}$$

Using our recipes, we can construct the above Hamiltonian as follows. We have taken the example of $N = 3$, $J = 1$, $B = 1$ and $r = 1$ under PBC.

```
ham_zz=hamobj.ham_zz(3,jz=1)
ham_z=hamobj.field_zham(3,mode='homogeneous',B=1)
ham=-ham_zz-ham_z
```

In the above program, the variable "ham" contains the required Hamiltonian matrix as shown in Eq. [6.15]

6.2.4.2   The homogeneous transverse Ising model in a non-zero uniform magnetic field

The Hamiltonian of the transverse field Ising model [127] is given by,

$$H = -J \sum_{i=1}^{N} \sigma_i^z \sigma_{i+1}^z - B \sum_{i=1}^{N} \sigma_i^x \qquad (6.16)$$

Using our recipes, we can construct the above Hamiltonian as follows. We have taken the example of $N = 3$, $J = 1$, $B = 1$, and $r = 1$ under PBC.

```
ham_zz=hamobj.ham_zz(3,jz=1)
ham_x=hamobj.field_xham(3,mode='homogeneous',B=1)
ham=-ham_zz-ham_x
```

In the above program, the variable ham contains the required Hamiltonian matrix as shown in Eq. [6.16]. The tilted field Ising model as well as other Ising type models with any neighbour interaction can simply be constructed suitably using similar prescriptions as given in Sections (6.2.4.1, 6.2.4.2).

6.3   The Heisenberg interaction

The one-dimensional Heisenberg model [117, 128–130] is an important model in physics, The Hamiltonian for the nearest-neighbour one-dimensional Heisenberg model for the case of N interacting spin half qubits is as follows

$$H = J \sum_{i=1}^{N} \vec{\sigma}_i . \vec{\sigma}_{i+1} = J \sum_{i=1}^{N} (\sigma_i^x \sigma_{i+1}^x + \sigma_i^y \sigma_{i+1}^y + \sigma_i^z \sigma_{i+1}^z) \qquad (6.17)$$

Note the dot product in the above Hamiltonian. If $J < 0$, the ground state of the above Hamiltonian is ferromagnetic and If $J > 0$, the ground state is antiferromagnetic. It can be solved analytically for its ground states as well as the excited states by Bethe Ansatz [128] for the case when $J < 0$ and by Hulthen's method [117] for the case when $J > 0$. We generalize the Heisenberg Hamiltonian to the rth neighbour with random interaction strengths using Eq. [6.17] as,

$$H = \sum_{i=1}^{N} J_r^i \vec{\sigma}_i . \vec{\sigma}_{i+r} = \sum_{i=1}^{N} J_r^i (\sigma_i^x \sigma_{i+r}^x + \sigma_i^y \sigma_{i+r}^y + \sigma_i^z \sigma_{i+r}^z), \qquad (6.18)$$

where J_r^i is the interaction strength which is also called the coupling constant between the the ith and $(i + r)^{th}$ spin. $\vec{\sigma} = (\sigma^x, \sigma^y, \sigma^z)$ where σ^j are the Pauli spin matrices, with $j = x$, y, z. Also note σ_{i+r} denotes the Pauli spin matrices corresponding to the rth neighbouring spin from σ_i. It is very easy to verify that

$$\vec{\sigma}_i . \vec{\sigma}_{i+r} = 2P_{i,i+r} - 1 \qquad (6.19)$$

Where $P_{i,i+r}$ is the permutation operator which swaps the ith and $(i+r)^{th}$ bit. We use the permutation operator to construct the Hamiltonian instead of using the naive form $\vec{\sigma}_i.\vec{\sigma}_{i+r}$.

```
heisenberg_hamiltonian(N,nn=1,mode='homogeneous',\
                       j=1.0,condition='periodic')
```

Parameters

In/Out	Argument	Description
[in]	N	It is the total number of spins
[in]	nn	nn stores the value of r describing the rth neighbour interaction of the Hamiltonian. The default value is equal to 1 (nearest neighbour interaction)
[in]	mode	It specifies whether the interaction strength is uniform or non-uniform If mode = 'homogeneous', then we have an uniform interaction ($J_r^i = J$). It is the default value If mode = 'inhomogeneous', then we have non-uniform interaction
[in]	j	If the mode = 'homogeneous', then j takes in a constant value, and the default value is 1 If the mode = 'inhomogeneous', then we pass a list holding the J_r^i value to parameter j. The ith entry represents the interaction strength between the ith and $(i+r)^{th}$ spin
[in]	condition	It defines the type of boundary condition if condition = 'periodic', periodic boundary conditions will be considered. It is the default value if condition = 'open', open boundary conditions will be considered
[out]	ham	It is an array of dimension (0:2**N−1,0:2**N−1), which is the required Hamiltonian

Implementation

```
def heisenberg_hamiltonian(self,N,nn=1,mode='homogeneous',j=1.0,\
                       condition='periodic'):
    """
    Construct Heisenberg interaction type Hamiltonian
    Input
        N: number of spins
        nn: specifying rth interaction
        mode: 'homogeneous' or 'inhomogeneous' magnetic field
        j: it list of values if mode='inhomogeneous', and constant if
            mode='homogeneous'
        condition:  defining boundary conditions of the Hamiltonian
```

```
Output
    ham: Hamiltonian

"""
# Checking entered boundary condition, PBC and OBC
assert condition == 'periodic' or condition == 'open'
# Number of interaction depending on the boundary condition
if condition == 'periodic':
    kn=N
else:
    kn=N-nn
# Only two modes are allowed, homogeneous and inhomogeneous
assert mode == 'homogeneous' or mode == 'inhomogeneous',\
"Entered mode is invalid"
```
Checking number of spins $N \geq 1$
```
assert N >= 1, "number of spins entered is not correct"
```
Checking valid r^{th} interaction
```
assert nn<=N-1 and nn>=1,"Not valid interaction"
ham=np.zeros([2**N,2**N],dtype='float64')
if mode == 'homogeneous':
```
        # Row index of matrix, $\sum_{i=1}^{N} J_r \vec{\sigma}_i.\vec{\sigma}_{i+r}$
```
    for i in range(0,2**N):
```
            # Column index of matrix, $\sum_{i=1}^{N} J_r \vec{\sigma}_i.\vec{\sigma}_{i+r}$
```
        for jj in range(0,2**N):
            col=qobj.decimal_binary(jj,N)
            for k in range(0,kn):
                k1=k+nn
                # For wrapping the Hamiltonian
                if k1>=N:
                    k1=k1-N
                col2=col.copy()
                temp=col2[k]
                col2[k]=col2[k1]
                col2[k1]=temp
                dec=qobj.binary_decimal(col2)
                if dec==i:
                    inn=1
                else:
                    inn=0
                ham[i,jj]=ham[i,jj]+(inn*j*2.0)
                if i==jj:
                    ham[i,jj]=ham[i,jj]-j
else:
```
    # Checking number of J_r^i interactions are valid according to BC
```
    if condition=='periodic':
        assert len(j)==N,\
        "Entered list of values j are not equal to number of
interaction in PBC"
    else:
        assert len(j)==N-nn,\
```

```
                    "Entered list of values j are not equal to number of
        interaction in OBC"
            # Row index of matrix, ∑_{i=1}^{N} J_r^i σ⃗_i·σ⃗_{i+r}
            for i in range(0,2**N):
                # Column index of matrix, ∑_{i=1}^{N} J_r^i σ⃗_i·σ⃗_{i+r}
                for jj in range(0,2**N):
                    col=qobj.decimal_binary(jj,N)
                    for k in range(0,kn):
                        k1=k+nn
                        # For wrapping Hamiltonian
                        if k1>=N:
                            k1=k1-N
                        col2=col.copy()
                        temp=col2[k]
                        col2[k]=col2[k1]
                        col2[k1]=temp
                        dec=qobj.binary_decimal(col2)
                        if dec==i:
                            inn=1
                        else:
                            inn=0
                        ham[i,jj]=ham[i,jj]+(inn*j[k]*2.0)
                        if i==jj:
                            ham[i,jj]=ham[i,jj]-j[k]
        return ham
```

Example

In this example, we are finding the Hamiltonian of the Heisenberg random interaction (Jr=(1,2,3)) with periodic boundary conditions.

```
ham_hie=hamobj.heisenberg_hamiltonian(3,mode='inhomogeneous',j=[1,2,3],
    condition='periodic')
print(ham_hie)
```

The preceding code prints the required Hamiltonian matrix.

```
[[ 6.  0.  0.  0.  0.  0.  0.  0.]
 [ 0. -4.  4.  0.  6.  0.  0.  0.]
 [ 0.  4.  0.  0.  2.  0.  0.  0.]
 [ 0.  0.  0. -2.  0.  2.  6.  0.]
 [ 0.  6.  2.  0. -2.  0.  0.  0.]
 [ 0.  0.  0.  2.  0.  0.  4.  0.]
 [ 0.  0.  0.  6.  0.  4. -4.  0.]
 [ 0.  0.  0.  0.  0.  0.  0.  6.]]
```

6.3.1 Variants of the Heisenberg model

The Hamiltonian described by Eq. [6.17] and Eq. [6.18] represent the Heisenberg XXX model. Of course, a generic Heisenberg model with nearest-neighbour interaction will be of the form,

$$H = \sum_{i=1}^{N}(J_{rx}\sigma_i^x\sigma_{i+r}^x + J_{ry}\sigma_i^y\sigma_{i+r}^y + J_{rz}\sigma_i^z\sigma_{i+r}^z), \tag{6.20}$$

here if $J_{rx} = J_{ry} = J_{rz}$, we call it the XXX model, if $J_{rx} \neq J_{ry} \neq J_{rz}$, it is called the XYZ model and if $J = J_{rx} = J_{ry} \neq J_{rz}$, it is called the XXZ model. These can be readily generalized to the rth neighbour models. Note that here J_{rj} does not contain superscript i as the original Heisenberg model does not have bond randomness.

6.3.1.1 The XXX model

Using our methods, we can construct the XXX Hamiltonian as follows. We have taken the example of $N = 3$, $J_{rx} = J_{ry} = J_{rz} = 1$ and $r = 1$ under PBC.

```
hamxx=hamobj.ham_xx(3,nn=1,mode='homogeneous',jx=1,condition='periodic')
hamyy=hamobj.ham_yy(3,nn=1,mode='homogeneous',jy=1,condition='periodic')
hamzz=hamobj.ham_zz(3,nn=1,mode='homogeneous',jz=1,condition='periodic')
ham=hamxx+hamyy+hamzz
```

In the above program, the variable "ham" contains the required Hamiltonian matrix of the XXX model.

6.3.1.2 The XYZ model

Using our methods, we can construct the XYZ Hamiltonian as follows. We have taken the example of $N = 3$, $J_{rx} = 0.7$, $J_{ry} = 0.8$, $J_{rz} = 0.9$ and $r = 1$ under PBC.

```
hamxx=hamobj.ham_xx(3,nn=1,mode='homogeneous',jx=0.7,condition='periodic')
hamyy=hamobj.ham_yy(3,nn=1,mode='homogeneous',jy=0.8,condition='periodic')
hamzz=hamobj.ham_zz(3,nn=1,mode='homogeneous',jz=0.9,condition='periodic')
ham=hamxx+hamyy+hamzz
```

In the above program, the variable "ham" contains the required Hamiltonian matrix of the XYZ model.

6.3.1.3 The Majumdar-Ghosh model

Hamiltonian of the Majumdar-Ghosh model [131–133] is given by,

$$H = J_1 \sum_{i=1}^{N}\vec{\sigma}_i.\vec{\sigma}_{i+1} + J_2 \sum_{i=1}^{N}\vec{\sigma}_i.\vec{\sigma}_{i+2} \tag{6.21}$$

Using our methods, we can construct the Majumdar-Ghosh Hamiltonian as follows. We have taken the example of $N = 6$, $J_1 = 1$, $J_2 = 0.5$ and $r = 1, 2$ under PBC. Note that J_1 and J_2 are homogeneous.

```
ham1=hamobj.heisenberg_hamiltonian(6,nn=1,mode='homogeneous',j=1.0,\
                    condition='periodic')
ham2=hamobj.heisenberg_hamiltonian(6,nn=2,mode='homogeneous',j=0.5,\
                    condition='periodic')
ham=ham1+ham2
```

In the above program, the variable "ham" contains the required Hamiltonian matrix of the Majumdar-Ghosh model.

6.4   The Dzyaloshinskii-Moriya interaction

The magnetic interactions between magnetic ions in a solid depend on numerous factors like neighbouring ions, temperature, external fields, etc. In some cases to describe the system one uses Hamiltonian involving simultaneous interaction between several spins also. Apart from the weak direct exchange coupling involving the dot product of the nearest neighbour spin operators $\vec{\sigma}_i.\vec{\sigma}_{i+1}$ in the Hamiltonian as discussed earlier in Section [6.3], there is, however, a small antisymmetric part due to the $L-S$ coupling between the spins or ions at site i and $i+1$. This favours spin canting (a tilt between the spins near to zero temperature), which we call an indirect coupling. It is interesting to note that this exists in a number of ionic solids, by which the short ranged exchange interaction between two non-neighbouring magnetic ions is mediated by a non-magnetic ion that resides between them in a phenomena similar to the Anderson super exchange [134]. This indirect coupling part of the Hamiltonian is called the Dzyaloshinskii-Moriya (DM) interaction [135, 136]. The Hamiltonian of the DM interaction is given by,

$$H = \sum_{i=1}^{N} \vec{D}.\vec{\sigma}_i \times \vec{\sigma}_{i+1}, \tag{6.22}$$

where $\vec{D} = (D_x, D_y, D_z)$ is the DM vector, whose components specify the coupling parameter of the DM interaction in the X, Y, Z directions, respectively, N is the total number of spins or qubits. Note that for the DM interaction between spins at sites i and $i+1$, the vector D is orthogonal to both the spins. That is, if the two interacting spins are lying in the plane of this book, then the DM vector is along the direction either into or out of the plane containing book. This cross product term as given in Eq. [6.22] is also related to what is known as the Heisenberg spin current [137]. Since the interaction is antisymmetric, the above Hamiltonian involves a cross product. We can recast Eq. [6.22] as,

$$H = \sum_{i=1}^{N} \begin{vmatrix} D_x & D_y & D_z \\ \sigma_i^x & \sigma_i^y & \sigma_i^z \\ \sigma_{i+1}^x & \sigma_{i+1}^y & \sigma_{i+1}^z \end{vmatrix}. \tag{6.23}$$

By expanding the determinant in equation Eq. [6.23] we get,

$$H = \sum_{i=1}^{N} \Big[D_x(\sigma_i^y \sigma_{i+1}^z - \sigma_i^z \sigma_{i+1}^y) + D_y(\sigma_i^z \sigma_{i+1}^x - \sigma_i^x \sigma_{i+1}^z)$$
$$+ D_z(\sigma_i^x \sigma_{i+1}^y - \sigma_i^y \sigma_{i+1}^x) \Big]. \tag{6.24}$$

We give the recipes for constructing these three terms in the Hamiltonian separately. The above formalism of antisymmetric exchange can be faithfully extended to r nearest neighbours also and in fact the recipes we give involve DM interaction between the spin at the site i and the spin at site the $i+r$ as

$$H = \sum_{i=1}^{N} \Big[D_{rx}^i(\sigma_i^y \sigma_{i+r}^z - \sigma_i^z \sigma_{i+r}^y) + D_{ry}^i(\sigma_i^z \sigma_{i+r}^x - \sigma_i^x \sigma_{i+r}^z)$$
$$+ D_{rz}^i(\sigma_i^x \sigma_{i+r}^y - \sigma_i^y \sigma_{i+r}^x) \Big]. \tag{6.25}$$

Also note that to accommodate the random couplings between i^{th} and $(i+r)^{th}$ spins, we make the components of $\vec{D}$ random by introducing the superscript i. It is noteworthy to mention that, in general, in Eq. [6.24], only one component of the vector D is non-zero, depending on the direction of the orientation of vector D.

6.4.1 DM vector in X direction

This corresponds to the first term in Eq. [6.25] which is,

$$H = \sum_{i=1}^{N} D_{rx}^{i}(\sigma_i^y \sigma_{i+r}^z - \sigma_i^z \sigma_{i+r}^y).$$ (6.26)

`dm_xham(N,nn=1,mode='homogeneous',dx=1.0, condition='periodic')`

Parameters

In/Out	Argument	Description
[in]	N	It is the total number of spins
[in]	nn	nn stores the value of r describing the rth neighbour interaction of the Hamiltonian. The default value is equal to 1 (nearest neighbour interaction)
[in]	mode	It specifies whether the interaction strength is uniform or non-uniform If mode = 'homogeneous', we have an uniform interaction ($D_{rx}^i = D$). It is the default value If mode = 'inhomogeneous', we have non-uniform interaction
[in]	dx	If the mode = 'homogeneous', dx takes in a constant value, and the default value is 1 If the mode = 'inhomogeneous', we pass a list holding the D_{rx}^i value to parameter dx. The ith entry represents the interaction strength between the ith and $(i+r)^{th}$ spin
[in]	condition	It defines the type of boundary condition if condition = 'periodic', periodic boundary conditions will be considered. It is the default value if condition = 'open', open boundary conditions will be considered
[out]	ham	It is an array of dimension $(0:2**N-1,0:2**N-1)$, which is the required Hamiltonian

Implementation

```
def dm_xham(self,N,nn=1,mode='homogeneous',dx=1.0, condition='periodic'):
    """

    Construct DM Hamiltonian in X direction
    Input
```

```
    N: number of spins
    nn: specifying rth interaction
    mode: 'homogeneous' or 'inhomogeneous' magnetic field
    dx: it list of values if mode='inhomogeneous', and constant if
        mode='homogeneous'
    condition:  defining boundary conditions of the Hamiltonian
Output
    ham: Hamiltonian

"""
# Checking entered boundary condition, PBC and OBC
assert condition == 'periodic' or condition == 'open'
# Number of interaction depending on the boundary condition
if condition == 'periodic':
    kn=N
else:
    kn=N-nn
# Only two modes are allowed, homogeneous and inhomogeneous
assert mode == 'homogeneous' or mode == 'inhomogeneous',\
"Entered mode is invalid"
# Checking number of spins $N \geq 1$
assert N >= 1, "number of spins entered is not correct"
# Checking valid $r^{th}$ interaction
assert nn<=N-1 and nn>=1,"Not valid interaction"
ham=np.zeros([2**N,2**N],dtype=np.complex_)
if mode == 'homogeneous':
    # Row index of matrix, $\sum_{i=1}^{N} D_{rx}(\sigma_i^y \sigma_{i+r}^z - \sigma_i^z \sigma_{i+r}^y)$
    for i in range(0,2**N):
        # Column index of matrix, $\sum_{i=1}^{N} D_{rx}(\sigma_i^y \sigma_{i+r}^z - \sigma_i^z \sigma_{i+r}^y)$
        for j in range(i,2**N):
            col=qobj.decimal_binary(j,N)
            for k in range(0,kn):
                k1=k+nn
                # For wrapping the Hamiltonian
                if k1>=N:
                    k1=k1-N
                col2=col.copy()
                col2[k]=1-col2[k]
                dec=qobj.binary_decimal(col2)
                inn1=complex(0.0,0.0)
                if dec==i:
                    phase=complex(0,1)*(-1)**col[k]
                    szpart=(-1)**col[k1]
                    inn1=phase*szpart
                col2=col.copy()
                col2[k1]=1-col2[k1]
                dec=qobj.binary_decimal(col2)
                inn2=complex(0.0,0.0)
                if dec==i:
                    phase=complex(0,1)*(-1)**col[k1]
```

```
                    szpart=(-1)**col[k]
                    inn2=phase*szpart
                ham[i,j]=ham[i,j]+((inn1-inn2)*dx)
            ham[j,i]=np.conjugate(ham[i,j])
    else:
        # Checking number of D_{rx}^i interactions are valid according to BC
        if condition=='periodic':
            assert len(dx)==N,\
            "The entered values of magnetic strengths are not equal to
number of spins"
        else:
            assert len(dx)==N-nn,\
            "The entered values of magnetic strengths are not equal to
number of spins"
        # Row index of matrix, \sum_{i=1}^{N} D_{rx}^i(\sigma_i^y\sigma_{i+r}^z - \sigma_i^z\sigma_{i+r}^y)
        for i in range(0,2**N):
            # Column index of matrix, \sum_{i=1}^{N} D_{rx}^i(\sigma_i^y\sigma_{i+r}^z - \sigma_i^z\sigma_{i+r}^y)
            for j in range(i,2**N):
                col=qobj.decimal_binary(j,N)
                for k in range(0,kn):
                    k1=k+nn
                    # For wrapping the Hamiltonian
                    if k1>=N:
                        k1=k1-N
                    col2=col.copy()
                    col2[k]=1-col2[k]
                    dec=qobj.binary_decimal(col2)
                    inn1=0.0
                    if dec==i:
                        phase=complex(0,1)*(-1)**col[k]
                        szpart=(-1)**col[k1]
                        inn1=phase*szpart
                    col2=col.copy()
                    col2[k1]=1-col2[k1]
                    dec=qobj.binary_decimal(col2)
                    inn2=0.0
                    if dec==i:
                        phase=complex(0,1)*(-1)**col[k1]
                        szpart=(-1)**col[k]
                        inn2=phase*szpart
                    ham[i,j]=ham[i,j]+((inn1-inn2)*dx[k])
                ham[j,i]=np.conjugate(ham[i,j])
    return ham
```

Example

In this example, we are finding the DM Hamiltonian with random interaction couplings
(Dr=(1,2,3)) in the X direction with periodic boundary condition.

```
ham=hamobj.dm_xham(3,mode='inhomogeneous',dx=[1,2,3],condition='periodic')
print(ham)
```

The preceding code prints the matrix elements of the required Hamiltonian as shown below.

```
[[0.-0.j 0.-1.j 0.-1.j 0.+0.j 0.+2.j 0.+0.j 0.+0.j 0.+0.j]
 [0.+1.j 0.-0.j 0.+0.j 0.+3.j 0.+0.j 0.-4.j 0.+0.j 0.+0.j]
 [0.+1.j 0.-0.j 0.-0.j 0.-5.j 0.+0.j 0.+0.j 0.+4.j 0.+0.j]
 [0.-0.j 0.-3.j 0.+5.j 0.-0.j 0.+0.j 0.+0.j 0.+0.j 0.-2.j]
 [0.-2.j 0.-0.j 0.-0.j 0.-0.j 0.-0.j 0.+5.j 0.-3.j 0.+0.j]
 [0.-0.j 0.+4.j 0.-0.j 0.-0.j 0.-5.j 0.-0.j 0.+0.j 0.+1.j]
 [0.-0.j 0.-0.j 0.-4.j 0.-0.j 0.+3.j 0.-0.j 0.-0.j 0.+1.j]
 [0.-0.j 0.-0.j 0.-0.j 0.+2.j 0.-0.j 0.-1.j 0.-1.j 0.-0.j]]
```

6.4.2 DM vector in Y direction

This corresponds to the second term in Eq. [6.25] which is,

$$H = \sum_{i=1}^{N} D_{ry}^i (\sigma_i^z \sigma_{i+r}^x - \sigma_i^x \sigma_{i+r}^z).$$

(6.27)

```
dm_yham(N,nn=1,mode='homogeneous',dy=1.0, condition='periodic')
```

Parameters

In/Out	Argument	Description
[in]	N	It is the total number of spins
[in]	nn	nn stores the value of r describing the rth neighbour interaction of the Hamiltonian. The default value is equal to 1 (nearest neighbour interaction)
[in]	mode	It specifies whether the interaction strength is uniform or non-uniform If mode = 'homogeneous', we have an uniform interaction ($D_{ry}^i = D$). It is the default value If mode = 'inhomogeneous', we have non-uniform interaction
[in]	dy	If the mode = 'homogeneous', dy takes in a constant value, and the default value is 1 If the mode = 'inhomogeneous', we pass a list holding the D_{ry}^i value to parameter dy. The ith entry represents the interaction strength between the ith and $(i + r)^{th}$ spin
[in]	condition	It defines the type of boundary condition if condition = 'periodic', periodic boundary conditions will be considered. It is the default value if condition = 'open', open boundary conditions will be considered
[out]	ham	It is an array of dimension (0:2**N−1,0:2**N−1) which is the required Hamiltonian

Implementation

```python
def dm_yham(self,N,nn=1,mode='homogeneous',dy=1.0, condition='periodic'):
    """

    Construct DM Hamiltonian in Y direction
    Input
        N: number of spins
        nn: specifying rth interaction
        mode: 'homogeneous' or 'inhomogeneous' magnetic field
        dy: it list of values if mode='inhomogeneous', and constant if
            mode='homogeneous'
        condition:  defining boundary conditions of the Hamiltonian
    Output
        ham: Hamiltonian
    """

    # Checking entered boundary condition, PBC and OBC
    assert condition == 'periodic' or condition == 'open'
    # Number of interaction depending on the boundary condition
    if condition == 'periodic':
        kn=N
    else:
        kn=N-nn
    # Only two modes are allowed, homogeneous and inhomogeneous
    assert mode == 'homogeneous' or mode == 'inhomogeneous',\
    "Entered mode is invalid"
    # Checking number of spins $N \geq 1$
    assert N >= 1, "number of spins entered is not correct"
    # Checking valid $r^{th}$ interaction
    assert nn<=N-1 and nn>=1,"Not valid interaction"
    ham=np.zeros([2**N,2**N],dtype='float64')
    if mode == 'homogeneous':
        # Row index of matrix, $\sum_{i=1}^{N} D_{ry}(\sigma_i^z \sigma_{i+r}^x - \sigma_i^x \sigma_{i+r}^z)$
        for i in range(0,2**N):
            # Column index of matrix, $\sum_{i=1}^{N} D_{ry}(\sigma_i^z \sigma_{i+r}^x - \sigma_i^x \sigma_{i+r}^z)$
            for j in range(i,2**N):
                col=qobj.decimal_binary(j,N)
                for k in range(0,kn):
                    k1=k+nn
                    # For wrapping the Hamiltonian
                    if k1>=N:
                        k1=k1-N
                    col2=col.copy()
                    col2[k1]=1-col2[k1]
                    dec=qobj.binary_decimal(col2)
                    inn1=0.0
                    if dec==i:
                        inn1=(-1)**col[k]
                    col2=col.copy()
                    col2[k]=1-col2[k]
                    dec=qobj.binary_decimal(col2)
                    inn2=0.0
```

```
                    if dec==i:
                        inn2=(-1)**col[k1]
                    ham[i,j]=ham[i,j]+((inn1-inn2)*dy)
                ham[j,i]=ham[i,j]
    else:
        # Checking number of D^i_ry interactions are valid according to BC
        if condition=='periodic':
            assert len(dy)==N,\
            "The entered values of magnetic strengths are not equal to
number of spins"
        else:
            assert len(dy)==N-nn,\
            "The entered values of magnetic strengths are not equal to
number of spins"
        # Row index of matrix, $\sum_{i=1}^{N} D^i_{ry}(\sigma^z_i \sigma^x_{i+r} - \sigma^x_i \sigma^z_{i+r})$
        for i in range(0,2**N):
            # Column index of matrix, $\sum_{i=1}^{N} D^i_{ry}(\sigma^z_i \sigma^x_{i+r} - \sigma^x_i \sigma^z_{i+r})$
            for j in range(i,2**N):
                col=qobj.decimal_binary(j,N)
                for k in range(0,kn):
                    k1=k+nn
                    # For wrapping the Hamiltonian
                    if k1>=N:
                        k1=k1-N
                    col2=col.copy()
                    col2[k1]=1-col2[k1]
                    dec=qobj.binary_decimal(col2)
                    inn1=0.0
                    if dec==i:
                        inn1=(-1)**col[k]
                    col2=col.copy()
                    col2[k]=1-col2[k]
                    dec=qobj.binary_decimal(col2)
                    inn2=0.0
                    if dec==i:
                        inn2=(-1)**col[k1]
                    ham[i,j]=ham[i,j]+((inn1-inn2)*dy[k])
                ham[j,i]=ham[i,j]
    return ham
```

Example

In this example, we are finding the DM Hamiltonian with random interaction couplings
(Dr=(1,1,1)) in the Y direction with periodic boundary condition.

```
ham=hamobj.dm_yham(3)
print(ham)
```

The preceding code prints the matrix elements of the required Hamiltonian as shown below.

```
[[ 0.  0.  0.  0.  0.  0.  0.  0.]
 [ 0.  0.  0.  2.  0. -2.  0.  0.]
```

```
[ 0.   0.   0.  -2.   0.   0.   2.   0.]
[ 0.   2.  -2.   0.   0.   0.   0.   0.]
[ 0.   0.   0.   0.   0.   2.  -2.   0.]
[ 0.  -2.   0.   0.   2.   0.   0.   0.]
[ 0.   0.   2.   0.  -2.   0.   0.   0.]
[ 0.   0.   0.   0.   0.   0.   0.   0.]]
```

6.4.3   DM vector in Z direction

This corresponds to the third term in Eq. [6.25] which is,

$$H = \sum_{i=1}^{N} D_{rz}^i (\sigma_i^x \sigma_{i+r}^y - \sigma_i^y \sigma_{i+r}^x). \tag{6.28}$$

```
dm_zham(N,nn=1,mode='homogeneous',dz=1.0, condition='periodic')
```

Parameters

In/Out	Argument	Description
[in]	N	It is the total number of spins
[in]	nn	nn stores the value of r describing the rth neighbour interaction of the Hamiltonian. The default value is equal to 1 (nearest neighbour interaction)
[in]	mode	It specifies whether the interaction strength is uniform or non-uniform If mode = 'homogeneous', we have an uniform interaction $(D_{rz}^i = D)$. It is the default value If mode = 'inhomogeneous', we have non-uniform interaction
[in]	dz	If the mode = 'homogeneous', dz takes in a constant value, and the default value is 1 If the mode = 'inhomogeneous', we pass a list holding the D_{rz}^i value to parameter dz. The ith entry represents the interaction strength between the ith and $(i+r)^{th}$ spin
[in]	condition	It defines the type of boundary condition if condition = 'periodic', periodic boundary conditions will be considered. It is the default value if condition = 'open', open boundary conditions will be considered
[out]	ham	It is an array of dimension $(0{:}2{**}N{-}1,0{:}2{**}N{-}1)$ which is the required Hamiltonian

Implementation

```
def dm_zham(self,N,nn=1,mode='homogeneous',dz=1.0, condition='periodic'):
    """
    Construct DM Hamiltonian in Z direction
    Input
        N: number of spins
        nn: specifying rth interaction
        mode: 'homogeneous' or 'inhomogeneous' magnetic field
        dz: it list of values if mode='inhomogeneous', and constant if
            mode='homogeneous'
        condition:  defining boundary conditions of the Hamiltonian
    Output
        ham: Hamiltonian

    """
    # Checking entered boundary condition, PBC and OBC
    assert condition == 'periodic' or condition == 'open'
    # Number of interaction depending on the boundary condition
    if condition == 'periodic':
        kn=N
    else:
        kn=N-nn
    # Only two modes are allowed, homogeneous and inhomogeneous
    assert mode == 'homogeneous' or mode == 'inhomogeneous',\
    "Entered mode is invalid"
    # Checking number of spins N >= 1
    assert N >= 1, "number of spins entered is not correct"
    # Checking valid rth interaction
    assert nn<=N-1 and nn>=1,"Not valid interaction"
    ham=np.zeros([2**N,2**N],dtype=np.complex_)
    if mode == 'homogeneous':
        # Row index of matrix, $\sum_{i=1}^{N} D_{rz}(\sigma_i^x \sigma_{i+r}^y - \sigma_i^y \sigma_{i+r}^x)$
        for i in range(0,2**N):
            # Column index of matrix, $\sum_{i=1}^{N} D_{rz}(\sigma_i^x \sigma_{i+r}^y - \sigma_i^y \sigma_{i+r}^x)$
            for j in range(i,2**N):
                col=qobj.decimal_binary(j,N)
                for k in range(0,kn):
                    k1=k+nn
                    # For wrapping the Hamiltonian
                    if k1>=N:
                        k1=k1-N
                    col2=col.copy()
                    col2[k]=1-col2[k]
                    col2[k1]=1-col2[k1]
                    dec=qobj.binary_decimal(col2)
                    inn1=complex(0.0,0.0)
                    inn2=complex(0.0,0.0)
                    if dec==i:
                        inn1=complex(0,1)*(-1)**col[k1]
                        inn2=complex(0,1)*(-1)**col[k]
```

```
                    ham[i,j]=ham[i,j]+((inn1-inn2)*dz)
                ham[j,i]=np.conjugate(ham[i,j])
        else:
            # Checking number of D_{rz}^i interactions are valid according to BC
            if condition=='periodic':
                assert len(dz)==N,\
                "The entered values of magnetic strengths are not equal to
        number of spins"
            else:
                assert len(dz)==N-nn,\
                "The entered values of magnetic strengths are not equal to
        number of spins"
            # Row index of matrix, \sum_{i=1}^{N} D_{rz}^i (\sigma_i^x \sigma_{i+r}^y - \sigma_i^y \sigma_{i+r}^x)
            for i in range(0,2**N):
                # Column index of matrix, \sum_{i=1}^{N} D_{rz}^i (\sigma_i^x \sigma_{i+r}^y - \sigma_i^y \sigma_{i+r}^x)
                for j in range(i,2**N):
                    col=qobj.decimal_binary(j,N)
                    for k in range(0,kn):
                        k1=k+nn
                        # For wrapping the Hamiltonian
                        if k1>=N:
                            k1=k1-N
                        col2=col.copy()
                        col2[k]=1-col2[k]
                        col2[k1]=1-col2[k1]
                        dec=qobj.binary_decimal(col2)
                        inn1=complex(0.0,0.0)
                        inn2=complex(0.0,0.0)
                        if dec==i:
                            inn1=complex(0,1)*(-1)**col[k1]
                            inn2=complex(0,1)*(-1)**col[k]
                        ham[i,j]=ham[i,j]+((inn1-inn2)*dz[k])
                    ham[j,i]=np.conjugate(ham[i,j])
    return ham
```

Example

In this example, we are finding the DM Hamiltonian with random interaction couplings
(Dr=(1,1,1)) in the Z direction with periodic boundary condition.

```
ham=hamobj.dm_zham(3)
print(ham)
```

The preceding code prints the matrix elements of the required Hamiltonian as shown below.

```
[[0.-0.j 0.+0.j 0.+0.j 0.+0.j 0.+0.j 0.+0.j 0.+0.j 0.+0.j]
 [0.-0.j 0.-0.j 0.+2.j 0.+0.j 0.-2.j 0.+0.j 0.+0.j 0.+0.j]
 [0.-0.j 0.-2.j 0.-0.j 0.+0.j 0.+2.j 0.+0.j 0.+0.j 0.+0.j]
 [0.-0.j 0.-0.j 0.-0.j 0.-0.j 0.+0.j 0.+2.j 0.-2.j 0.+0.j]
 [0.-0.j 0.+2.j 0.-2.j 0.-0.j 0.-0.j 0.+0.j 0.+0.j 0.+0.j]
 [0.-0.j 0.-0.j 0.-0.j 0.-2.j 0.-0.j 0.-0.j 0.+2.j 0.+0.j]
 [0.-0.j 0.-0.j 0.-0.j 0.+2.j 0.-0.j 0.-2.j 0.-0.j 0.+0.j]
 [0.-0.j 0.-0.j 0.-0.j 0.-0.j 0.-0.j 0.-0.j 0.-0.j 0.-0.j]]
```

6.4.4 An Illustrative example of constructing a combination Hamiltonian

Let us see how to construct a combination Hamiltonian which involves the Hamiltonian of the Heisenberg model with DM interaction. In the following example, we have chosen the direction of the DM vector to be along Z.

$$H = J \sum_{i=1}^{N} \vec{\sigma}_i . \vec{\sigma}_{i+1} + D_z \sum_{i=1}^{N} (\sigma_i^x \sigma_{i+1}^y - \sigma_i^y \sigma_{i+1}^x). \tag{6.29}$$

Using our codes in this chapter, we can construct the above Hamiltonian as follows. We have taken the example of $N = 3$, $J = 1$, $D_z = 1$ and $r = 1$ under PBC.

```
hamz=hamobj.dm_zham(3)
ham_hie=hamobj.heisenberg_hamiltonian(3)
ham=ham_hie+hamz
```

In the above program, the variable 'ham' contains the required Hamiltonian matrix.

7

Generating Random Matrices and Random Vectors

In this chapter we give the numerical recipes to construct random numbers following a particular distribution and also construct random states and density matrices, here we have used the Gaussian and the uniform distribution. These are the most important distributions used in constructing random quantum states [138] as we progress through the chapter. Moreover, there are recipes to construct random matrices which occur in random matrix theory [139–141] and boderlining areas of quantum information. At the end of the chapter, recipes to construct the random density matrix are also given which will be very useful for people working on random quantum mixed states [142, 143]. The examples indicating the output using these recipes are not given for a specific reason that everytime the user prints an output, it will be different as they are from random processes. In this chapter we will be primarily discussing the class, **QRandom**. The class is written inside the Python module **chap7_randommatrix.py**. The class **QRandom** contains all the methods pertaining to generating random matrices from the uniform and Gaussian distribution. To import the class and create its objects, you have to write the following code,

```
# Importing the QRandom class
from QuantumInformation import QRandom

# Instantiate the object of the QRandom class
qrobj=QRandom()
```

7.1 Random number generator for Gaussian distribution

The probability density function for a Gaussian distribution is given by,

$$f(x) = \frac{1}{\sigma\sqrt{2\pi}} e^{-\frac{1}{2}\left(\frac{x-\mu}{\sigma}\right)^2} = \mathcal{N}(\mu, \sigma^2), \tag{7.1}$$

where μ, σ^2 are the mean and variance of the probability distribution, respectively. In Python, we use Numpy's random number routine to generate numbers from uniform or Gaussian distribution.

7.1.1 For a real random matrix whose elements are chosen from the Gaussian distribution

```
random_gaussian_rvec(tup,mu=0,sigma=1)
```

DOI: 10.1201/9781003285489-7

Parameters

In/Out	Argument	Description
[in]	tup	It holds the information of the dimension of the output matrix
[in]	mu	mu is the mean of the distribution. Its default value is zero
[in]	sigma	sigma is the standard deviation of the distribution. Its default value is 1
[out]	gauss_state	It is the array of dimensions assigned in the variable tup, and it is a random real matrix

Implementation

```
def random_gaussian_rvec(self,tup,mu=0,sigma=1):
    """
    Construct a real random state from a Gaussian distribution
    Input:
        tup: tuple holds the dimension of the output matrix.
        mu: it is the average value of the gaussian distribution.
        sigma: it is the standard deviation of the gaussian distribution
    Output:
        gauss_state: it the gaussian distributed real state
    """
    gauss_state=np.random.normal(mu, sigma, tup)
    return gauss_state
```

Example

```
state_gauss=qrobj.random_gaussian_rvec((3,3))
```

In the above example, **state_gauss** is a 3×3 real matrix which stores the random numbers from the Gaussian distribution with zero mean and unit standard deviation.

7.1.2　For a complex random matrix whose real and imaginary parts are individually chosen from the Gaussian distribution

```
random_gaussian_cvec(tup,mu=0,sigma=1)
```

Parameters

In/Out	Argument	Description
[in]	tup	It holds the information of the dimension of the output matrix
[in]	mu	mu is the mean of the distribution
[in]	sigma	sigma is the standard deviation of the distribution
[out]	state_gauss	It is the array of dimensions assigned in the variable tup, and it is a random complex matrix

Implementation

```
def random_gaussian_cvec(self,tup,mu=0,sigma=1):
    """
    Construct a complex random state from a Gaussian distribution
    Input:
        tup: tuple holds the dimension of the output matrix.
        mu: it is the average value of the gaussian distribution.
        sigma: it is the standard deviation of the gaussian distribution
    Output:
        gauss_state: it the gaussian distributed complex state
    """
    # Real part selected from Gaussian distribution
    rpart=np.random.normal(mu, sigma, tup)
    # Complex part selected from Gaussian distribution
    cpart=np.random.normal(mu, sigma, tup)
    state_gauss=np.zeros(tup,dtype=np.complex_)
    state_gauss=rpart+(complex(0,1)*cpart)
    return state_gauss
```

Example

```
state_gauss=qrobj.random_gaussian_cvec((3,3))
```

In the above example, state_gauss is a 3×3 complex matrix which stores the random numbers from the Gaussian distribution with zero mean and unit standard deviation.

7.2 Random number generator for uniform distribution in the range (a,b)

The uniform random number generator generates a random number in the arbitrary range (a, b). The following method generates a matrix whose matrix elements are sampled from the uniform distribution in the range (a,b).

7.2.1 For a real random matrix whose elements are chosen from uniform distribution

random_unifrom_rvec(tup,low=0.0,high=1.0)

Parameters

In/Out	Argument	Description
[in]	tup	It holds the information of the dimension of the output matrix
[in]	low	It is the lowest value of the range. Its default value is 0
[in]	high	It is the highest value of the range. Its default value is equal to 1
[out]	uniform_state	It is the array of dimensions assigned in the variable tup, and it is a random real matrix

Implementation

```
def random_unifrom_rvec(self,tup,low=0.0,high=1.0):
    """
    Construct a real random state from a uniform distribution
    Input:
        tup: tuple holds the dimension of the output matrix.
        low: it is the lower bound of the uniform distribution.
        high: it is the upper bound of the uniform distribution
    Output:
        uniform_state: it the uniform distributed real state
    """
    uniform_state=np.random.uniform(low=low, high=high, size=tup)
    return uniform_state
```

Example

state_uniform=qrobj.random_unifrom_rvec((3,3),low=1,high=2)

In the above example, state_uniform is a 3×3 real matrix which stores the random numbers chosen from a uniform distribution in the range (1,2).

7.2.2 For a complex random matrix whose elements are chosen from uniform distribution

random_unifrom_cvec(tup,low=0.0,high=1.0)

Parameters

In/Out	Argument	Description
[in]	tup	It holds the information of the dimension of the output matrix
[in]	low	It is the lowest value of the range. Its default value is 0
[in]	high	It is the highest value of the range. Its default value is equal to 1
[out]	uniform_state	It is the array of dimensions assigned in the variable tup, and it is a random complex matrix

Implementation

```python
def random_unifrom_cvec(self,tup,low=0.0,high=1.0):
    """
    Construct a complex random state from a uniform distribution
    Input:
        tup: tuple holds the dimension of the output matrix.
        low: it is the lower bound of the uniform distribution.
        high: it is the upper bound of the uniform distribution
    Output:
        uniform_state: it the uniform distributed complex state
    """
    # Real part chosen from uniform distribution
    uniform_rpart=np.random.uniform(low=low, high=high, size=tup)
    # Complex part chosen from uniform distribution
    uniform_cpart=np.random.uniform(low=low, high=high, size=tup)
    uniform_state=uniform_rpart+(complex(0,1)*uniform_cpart)
    return uniform_state
```

Example

```python
state_uniform=qrobj.random_unifrom_cvec((2,2),low=-1,high=1)
```

In the above example, state_uniform is a 2×2 complex matrix which stores the random numbers chosen from a uniform distribution in the range $(-1, 1)$.

7.3   Random real symmetric matrices

Any random real symmetric matrix B can be generated from any general matrix A whose elements are from a Gaussian or a uniform distribution as follows,

$$B = \frac{1}{2}(A + A^T) \qquad (7.2)$$

```
random_symmetric_matrix(size,distribution="gaussian",mu=0,\
                        sigma=1,low=0.0,high=1.0)
```

Parameters

In/Out	Argument	Description
[in]	size	It holds the information of the dimension of the output matrix
[in]	distribution	It holds a string value defining the type of distribution. if distribution = 'gaussian', Gaussian distribution will be chosen. It is the default value ; if distribution = 'uniform', uniform distribution will be chosen
[in]	low	If distribution = 'uniform', low (integer) is the lowest value of the range. The default value is 0 ; If distribution = 'gaussian', low is not referenced
[in]	high	If distribution = 'uniform', high (integer) is the highest value of the range. The default value is 1; If distribution = 'gaussian', b is not referenced
[in]	mu	If dist = 'gaussian', mu is the mean of the distribution; If distribution = 'uniform', mu is not referenced
[in]	sigma	If distribution = 'gaussian', sigma is the standard deviation of the distribution; If distribution = 'uniform', sigma is not referenced
[out]	smatrix	It is the array of dimensions assigned in the variable size, and it is a random real symmetric matrix

Implementation

```
def random_symmetric_matrix(self,size,distribution="gaussian",mu=0,\
                            sigma=1,low=0.0,high=1.0):
    """
    Construct a random symmetric matrix
    Input:
        size: it is a tuple (a,b), where a in number of rows and b is
              number of columns of the symmetric matrix
        distribution: it is the type of distribution
    Output:
        smatrix: symmetric matrix
    """
```

```
# Checking whether entered mode for distribution is correct or not
assert re.findall("^gaussian|^uniform",distribution),\
"Invalid distribution type"
if distribution == 'gaussian':
    smatrix=np.random.normal(mu, sigma, size=size)
if distribution == 'uniform':
    smatrix=np.random.uniform(low=low, high=high, size=size)
smatrix=smatrix+np.matrix.transpose(smatrix)
smatrix=0.5*smatrix
return smatrix
```

Example

```
random_smatrix=qrobj.random_symmetric_matrix(size=(3,3),\
                                    distribution='gaussian',\
                                    mu=2,sigma=0.5)
```

In the above example, random_smatrix is a 3×3 real symmetric matrix which stores the random numbers selectively chosen from the Gaussian distribution with mean 2 and standard deviation 0.5.

7.4   Random complex Hermitian matrices

Any random complex Hermitian matrix B can be generated from any complex general matrix A whose elements are from a Gaussian or a uniform distribution as follows,

$$B = \frac{1}{2}(A + A^{\dagger}) \tag{7.3}$$

```
random_hermitian_matrix(self,size,distribution="gaussian",mu=0,\
                        sigma=1,low=0.0,high=1.0)
```

Parameters

In/Out	Argument	Description
[in]	size	It holds the information of the dimension of the output matrix
[in]	distribution	It holds a string value defining the type of distribution. if distribution = 'gaussian', Gaussian distribution will be chosen. It is the default value ; if distribution = 'uniform', uniform distribution will be chosen
[in]	low	If distribution = 'uniform', low (integer) is the lowest value of the range. The default value is 0 ; If distribution = 'gaussian', low is not referenced
[in]	high	If distribution = 'uniform', high (integer) is the highest value of the range. The default value is 1; If distribution = 'gaussian', b is not referenced

In/Out	Argument	Description
[in]	mu	If dist = 'gaussian', mu is the mean of the distribution; If distribution = 'uniform', mu is not referenced
[in]	sigma	If distribution = 'gaussian', sigma is the standard deviation of the distribution; If distribution = 'uniform', sigma is not referenced
[out]	hmatrix	It is the array of dimensions assigned in the variable size, and it is a random complex Hermitian matrix

Implementation

```
def random_hermitian_matrix(self,size,distribution="gaussian",mu=0,\
                            sigma=1,low=0.0,high=1.0):
    """
    Construct a random Hermitian matrix
    Input:
        size: it is a tuple (a,b), where a in number of rows and b is
              number of columns of the hermitian matrix
        distribution: it is the type of distribution
    Output:
        hmatrix: hermitian matrix
    """
    # Checking whether entered mode for distribution is correct or not
```

```
    assert re.findall("^gaussian|^uniform",distribution),\
    "Invalid distribution type"
    hmatrix=np.zeros([size[0],size[1]],dtype=np.complex_)
    if distribution=='gaussian':
        for i in range(0,hmatrix.shape[0]):
            for j in range(0,hmatrix.shape[1]):
                hmatrix[i,j]=complex(np.random.normal(mu, sigma, size= None
),\
                        np.random.normal(mu, sigma, size= None))
    else:
        for i in range(0,hmatrix.shape[0]):
            for j in range(0,hmatrix.shape[1]):
                hmatrix[i,j]=complex(np.random.uniform(low=low, high=high,\
                        size=None),np.random.uniform(low=low, high=high,\
                                size=None))
    hmatrix=hmatrix+np.matrix.conjugate(np.matrix.transpose(hmatrix))
    hmatrix=0.5*hmatrix
    return hmatrix
```

Example

```
random_hmatrix=qrobj.random_hermitian_matrix(size=(3,3),\
                                    distribution='uniform',\
                                    low=5,high=10)
```

In the above example, random_hmatrix is a 3×3 complex Hermitian matrix which stores the random numbers selectively chosen from a uniform distribution in the range (5,10).

7.5 Random unitary matrices

For generating a random unitary matrix B, generate any general real random matrix A with elements chosen from a Gaussian or a uniform distribution. Now, Perform a QR decomposition on the matrix A which will give us a random unitary matrix B such that $BB^\dagger = B^\dagger B = \mathbb{I}$, in the case of B being a real matrix, then we will be generating a random orthogonal matrix.

7.5.1 Random real orthogonal matrix

```
random_orthogonal_matrix(size,distribution="gaussian",mu=0,\
                        sigma=1,low=0.0,high=1.0)
```

Parameters

In/Out	Argument	Description
[in]	size	It holds the information of the dimension of the output matrix
[in]	distribution	It holds a string value defining the type of distribution. if distribution = 'gaussian', Gaussian distribution will be chosen. It is the default value ; if distribution = 'uniform', uniform distribution will be chosen
[in]	low	If distribution = 'uniform', low (integer) is the lowest value of the range. The default value is 0 ; If distribution = 'gaussian', low is not referenced
[in]	high	If distribution = 'uniform', high (integer) is the highest value of the range. The default value is 1; If distribution = 'gaussian', b is not referenced
[in]	mu	If dist = 'gaussian', mu is the mean of the distribution; If distribution = 'uniform', mu is not referenced
[in]	sigma	If distribution = 'gaussian', sigma is the standard deviation of the distribution; If distribution = 'uniform', sigma is not referenced
[out]	omatrix	It is the array of dimensions assigned in the variable size, and it is a random real orthogonal matrix

Implementation

```
def random_orthogonal_matrix(self,size,distribution="gaussian",mu=0,\
                    sigma=1,low=0.0,high=1.0):
    """
    Construct a random orthogonal matrix
    Input:
        size: it is a tuple (a,b), where a in number of rows and b is
              number of columns of the orthogonal matrix
        distribution: it is the type of distribution
    Output:
        omatrix: orthogonal matrix
    """
    # Checking whether entered mode for distribution is correct or not
    assert re.findall("^gaussian|^uniform",distribution),\
    "Invalid distribution type"

    if distribution == 'gaussian':
```

```
          omatrix=np.random.normal(mu, sigma, size= size)
      else:
          omatrix=np.random.uniform(low=low, high=high, size=size)
      omatrix,r=np.linalg.qr(omatrix,mode='complete')
      return omatrix
```

Example

```
random_omatrix=qrobj.random_orthogonal_matrix(size=(3,3),\
                                        distribution='uniform',\
                                        low=2,high=3)
```

In the above example, random_omatrix is a 3×3 real orthogonal matrix which stores the random numbers selectively chosen from the uniform distribution chosen from the range (2,3).

7.5.2   Random complex unitary matrix

```
random_unitary_matrix(size,distribution="gaussian",mu=0,\
                      sigma=1,low=0.0,high=1.0)
```

Parameters

In/Out	Argument	Description
[in]	size	It holds the information of the dimension of the output matrix
[in]	distribution	It holds a string value defining the type of distribution. if distribution = 'gaussian', Gaussian distribution will be chosen. It is the default value ; if distribution = 'uniform', uniform distribution will be chosen
[in]	low	If distribution = 'uniform', low (integer) is the lowest value of the range. The default value is 0 ; If distribution = 'gaussian', low is not referenced

In/Out	Argument	Description
[in]	high	If distribution = 'uniform', high (integer) is the highest value of the range. The default value is 1; If distribution = 'gaussian', b is not referenced
[in]	mu	If dist = 'gaussian', mu is the mean of the distribution; If distribution = 'uniform', mu is not referenced
[in]	sigma	If distribution = 'gaussian', sigma is the standard deviation of the distribution; If distribution = 'uniform', sigma is not referenced
[out]	umatrix	It is the array of dimensions assigned in the variable size, and it is a random complex unitary matrix

Implementation

```python
def random_unitary_matrix(self,size,distribution="gaussian",mu=0,\
                          sigma=1,low=0.0,high=1.0):
    """
    Construct a unitary matrix
    Input:
        size: it is a tuple (a,b), where a in number of rows and b is
              number of columns of the unitary matrix
        distribution: it is the type of distribution
    Output:
        umatrix: unitary matrix
    """
    # Checking whether entered mode for distribution is correct or not
    assert re.findall("^gaussian|^uniform",distribution),\
    "Invalid distribution type"
    umatrix=np.zeros([size[0],size[1]],dtype=np.complex_)

    if distribution == 'gaussian':
        for i in range(0,umatrix.shape[0]):
            for j in range(0,umatrix.shape[1]):
                umatrix[i,j]=complex(np.random.normal(mu,sigma,size=None),\
                        np.random.normal(mu,sigma,size=None))
    else:
        for i in range(0,umatrix.shape[0]):
            for j in range(0,umatrix.shape[1]):
                umatrix[i,j]=complex(\
                        np.random.uniform(low=low,high=high,size=None),\
                        np.random.uniform(low=low,high=high,size=None))
    umatrix,r=np.linalg.qr(umatrix,mode='complete')
    return umatrix
```

Example

```
random_umatrix=qrobj.random_unitary_matrix(size=(3,3),distribution='
    gaussian')
```

In the above example, random_umatrix is a 3×3 complex unitary matrix which stores the random numbers selectively chosen from a Gaussian distribution of zero mean and unit variance.

7.6   Real Ginibre matrix

The real Ginibre matrix is a random matrix with the elements chosen from an independent and identically distributed random variables chosen from the normal distribution $\mathcal{N}(0,1)$.

```
random_real_ginibre(N)
```

Parameters

In/Out	Argument	Description
[in]	N	It is the dimension (N×N) of the output matrix
[out]	state	It is the array holding the real Ginibre matrix

Implementation

```
def random_real_ginibre(self,N):
    """
    Construct a real Ginibre matrix
    Input:
        N: dimension of the NxN Ginibre matrix
    Output:
        real ginibre matrix
    """
    return np.random.normal(0.0, 1.0, size= (N,N))
```

Example

```
random_ginibre=qrobj.random_real_ginibre(2)
```

In the above example, random_ginibre is the 2×2 required real Ginibre matrix.

7.7 Complex Ginibre matrix

The complex Ginibre matrix is a random matrix with each of the real and imaginary elements individually chosen from an independent and identically distributed random variables chosen from the normal distribution $\mathcal{N}(0, \frac{1}{2})$.

```
random_complex_ginibre(N)
```

Parameters

In/Out	Argument	Description
[in]	N	It is the dimension (N×N) of the output matrix
[out]	cginibre	It is the array holding the complex Ginibre matrix

Implementation

```python
def random_complex_ginibre(self,N):
    """

    Construct a complex Ginibre matrix
    Input:
        N: dimension of the NxN complex Ginibre
    Output:
        cginibre: complex ginibre matrix
    """
    cginibre=np.zeros([N,N],dtype=np.complex_)
    for i in range(0,cginibre.shape[0]):
        for j in range(0,cginibre.shape[1]):
            cginibre[i,j]=complex(np.random.normal(0.0, 1.0, size= None),\
                np.random.normal(0.0, 1.0, size= None))
    return cginibre
```

Example

```
random_ginibre=qrobj.random_complex_ginibre(2)
```

In the above example, random_ginibre is the 2×2 required complex Ginibre matrix.

7.8 Wishart matrices

Wishart matrices are positive definite matrices which can be constructed using the Ginibre matrix G as follows.

$$W = GG^{\dagger} \tag{7.4}$$

7.8.1 Real Wishart matrix

```
random_real_wishart(N)
```

Parameters

In/Out	Argument	Description
[in]	N	It is the dimension (N×N) of the output matrix
[out]	rwishart	It is the array holding the real Wishart matrix

Implementation

```python
def random_real_wishart(self,N):
    """
    Construct a real Wishart matrix
    Input:
        N: dimension of the NxN real Wishart matrix
    Output:
        rwishart: real Wishart matrix
    """
    g=self.random_real_ginibre(N)
    rwishart=np.matmul(g,np.matrix.transpose(g))
    return rwishart
```

Example

```
random_wishart=qrobj.random_real_wishart(2)
```

In the above example, random_wishart is the 2×2 required real Wishart matrix.

7.8.2 Complex Wishart matrix

```
random_complex_wishart(N)
```

Parameters

In/Out	Argument	Description
[in]	N	It is the dimension (N×N) of the output matrix
[out]	cwishart	It is the array holding the complex Wishart matrix

Implementation

```
def random_complex_wishart(self,N):
    """
    Input:
        N: dimension of the NxN complex Wishart matrix
    Output:
        cwishart: complex Wishart matrix
    """
    g=self.random_complex_ginibre(N)
    cwishart=np.matmul(g,np.matrix.conjugate(np.matrix.transpose(g)))
    return cwishart
```

Example

```
random_wishart=qrobj.random_complex_wishart(3)
```

In the above example, random_wishart is the 3×3 required complex Wishart matrix.

7.9   Random probability vector

A random probability vector whose entries are probabilities can be given by,

$$P = (p_1, p_2, \cdots, p_n) \tag{7.5}$$

here, $0 \leq p_i \leq 1$, also the p_i's are chosen from a uniform distribution between the range (0,1). Note that, we have to normalize the vector P such that $\sum_{i=1}^{n} p_i = 1$.

Parameters

```
subroutine RANDPVEC(d,state)
```

In/Out	Argument	Description
[in]	N	It is dimension of the probability vector
[out]	prob_vec	It is array of dimension (0:N−1) which is the column vector containing the probabilities

Implementation

```
def random_probability_vec(self,N):
    """

    Constructs a random probability vector
    Input:
        N: dimension of the probability vector.
    Output:
        prob_vec: The probability vector
    """
    prob_vec=np.random.uniform(low=0,high=1.0,size=N)
    norm=prob_vec.sum()
    prob_vec=prob_vec/norm
    return prob_vec
```

Example

```
pvec=qrobj.random_probability_vec(5)
```

In the above example, pvec is the 5×1 required probability vector which is normalized.

7.10   Random pure quantum state vector

The random quantum state vector [138, 144] of dimension n can be described as,

$$|\psi\rangle = \sum_{j=1}^{n} c_j |j\rangle, \tag{7.6}$$

where c_j's are the random numbers, either real or complex chosen from any random distribution, here we use the uniform and Gaussian distribution, such that $\sum_{j=1}^{n} |c_j|^2 = 1$. Note that, $|j\rangle$ is the computational basis. Random quantum states are solely defined analytically based on the normalization condition only which is identical to the equation of a n-dimensional unit sphere. This probability distribution is proportional to the Dirac delta function such that the $P(c_1, c_2, \ldots, c_n) \propto \delta(\sum_{j=1}^{n} |c_j|^2 - 1)$.

7.10.1   Random real state vector

```
random_qrstate(N,distribution='gaussian',\
                mu=0,sigma=1,low=0.0,high=1.0)
```

Parameters

In/Out	Argument	Description
[in]	N	It holds the value of total number of qubits
[in]	distribution	It holds a string value defining the type of distribution. if distribution = 'gaussian', Gaussian distribution will be chosen. It is the default value ; if distribution = 'uniform', uniform distribution will be chosen
[in]	low	If distribution = 'uniform', low (integer) is the lowest value of the range. The default value is 0 ; If distribution = 'gaussian', low is not referenced
[in]	high	If distribution = 'uniform', high (integer) is the highest value of the range. The default value is 1; If distribution = 'gaussian', b is not referenced
[in]	mu	If dist = 'gaussian', mu is the mean of the distribution; If distribution = 'uniform', mu is not referenced
[in]	sigma	If distribution = 'gaussian', sigma is the standard deviation of the distribution; If distribution = 'uniform', sigma is not referenced
[out]	qrstate	It is an array which holds a random real quantum state

Implementation

```
def random_qrstate(self,N,distribution='gaussian',\
                mu=0,sigma=1,low=0.0,high=1.0):
    """
    Constructs a random real pure quantum state
    Input:
        N: number of qubits
        distribution: it is the type of distribution
    Output:
        qrstate: real quantum state
    """
    assert re.findall("^gaussian|^uniform",distribution),\
    "Invalid distribution type"
    if distribution=='gaussian':
        qrstate=self.random_gaussian_rvec(2**N,mu=0,sigma=1)
    else:
        qrstate=self.random_unifrom_rvec(2**N,low=0.0,high=1.0)
    norm=np.matmul(np.matrix.transpose(qrstate),qrstate)
    qrstate=qrstate/np.sqrt(norm)
    return qrstate
```

Example

```
rstate=qrobj.random_qrstate(4,distribution='uniform',\
                    low=0.0,high=10.0)
```

In the above example, rstate is the $2^4 \times 1$ real random quantum state vector whose entries are selectively chosen from the uniform distribution, in the range (0,10).

7.10.2 Random pure complex state vector

```
random_qcstate(self,N,distribution='gaussian',\
                mu=0,sigma=1,low=0.0,high=1.0)
```

Parameters

In/Out	Argument	Description
[in]	N	It holds the value of total number of qubits
[in]	distribution	It holds a string value defining the type of distribution. if distribution = 'gaussian', Gaussian distribution will be chosen. It is the default value ; if distribution = 'uniform', uniform distribution will be chosen
[in]	low	If distribution = 'uniform', low (integer) is the lowest value of the range. The default value is 0 ; If distribution = 'gaussian', low is not referenced
[in]	high	If distribution = 'uniform', high (integer) is the highest value of the range. The default value is 1; If distribution = 'gaussian', b is not referenced
[in]	mu	If dist = 'gaussian', mu is the mean of the distribution; If distribution = 'uniform', mu is not referenced
[in]	sigma	If distribution = 'gaussian', sigma is the standard deviation of the distribution; If distribution = 'uniform', sigma is not referenced
[out]	qcstate	It is an array which holds a random complex quantum state

Implementation

```
def random_qcstate(self,N,distribution='gaussian',\
                mu=0,sigma=1,low=0.0,high=1.0):
    """
    Constructs a random complex pure quantum state
```

```
Input:
    N: number of qubits
    distribution: it is the type of distribution
Output:
    qcstate: complex quantum state
"""
assert re.findall("^gaussian|^uniform",distribution),\
"Invalid distribution type"
if distribution=='gaussian':
    qcstate=self.random_gaussian_cvec(2**N,mu=0,sigma=1)
else:
    qcstate=self.random_unifrom_cvec(2**N,low=0.0,high=1.0)
norm=abs(np.matmul(\
                np.matrix.conjugate(\
                                np.matrix.transpose(qcstate)),\
                                qcstate))
qcstate=qcstate/np.sqrt(norm)
return qcstate
```

Example

```
cstate=qrobj.random_qcstate(4,distribution='gaussian',\
                        mu=2,sigma=1)
```

In the above example, cstate is the $2^4 \times 1$ complex random pure quantum state vector whose elements are selectively chosen from the Gaussian distribution with mean equal to 2 and standard deviation equal to 1.

7.11   Random density matrices

A random density matrix [141, 142] ρ can be generated using a real or a complex Ginibre matrix G as follows,

$$\rho = \frac{GG^\dagger}{Tr(GG^\dagger)} \tag{7.7}$$

The matrix ρ is by construction Hermitian, positive definite and normalized. The ensemble so generated is called the Hilbert–Schmidt ensemble.

7.11.1   Random real density matrix

```
subroutine RANDDMR(d,state)
```

Parameters

In/Out	Argument	Description
[in]	N	It is the number of qubits
[out]	rden	It is array of dimension $(0:2^N-1,0:2^N-1)$ which is the random real density matrix

Implementation

```
def random_rden(self,N):
    """

    Constructs a random real pure density matrix
    Input:
        N: number of qubits
    Output:
        rden: real random density matrix
    """
    rden=self.random_real_wishart(2**N)
    rden=rden/np.trace(rden)
    return rden
```

Example

```
rden=qrobj.random_rden(4)
```

In the above example, the rden variable holds a $2^4 \times 2^4$ real density matrix.

7.11.2   Random complex density matrix

```
subroutine RANDDMC(d,state)
```

Parameters

In/Out	Argument	Description
[in]	N	It is the number of qubits
[out]	cden	It is array of dimension $(0:2^N-1,0:2^N-1)$ which is the random complex density matrix

Implementation

```
def random_cden(self,N):
    """

    Constructs a random complex pure density matrix
    Input:
```

```
    N: number of qubits
Output:
    cden: complex random density matrix
"""
cden=self.random_complex_wishart(2**N)
cden=cden/np.trace(cden)
return cden
```

Example

```
cden=qrobj.random_cden(4)
```

In the above example, the cden variable holds a $2^4 \times 2^4$ complex density matrix.

Bibliography

[1] B. Venners, The making of Python: A conversation with guido van rossum, part i, Artima Developer (Jan. 2003). url: http://www. artima. com/intv/pythonP. html (2003).

[2] A. Scopatz and K. D. Huff, *Effective computation in physics: field guide to research with Python* ("O'Reilly Media, Inc." 2015).

[3] Y. Kanetkar, *Let us C* (BPB publications 2018).

[4] Y. P. Kanetkar, *Let us C++* (BPB Publications 1999).

[5] S. J. Chapman, *Fortran 90/95 for scientists and engineers* (McGraw-Hill Higher Education 2004).

[6] M. S. Ramkarthik and P. D. Solanki, *Numerical Recipes in Quantum Information Theory and Quantum Computing: An Adventure in Fortran 90* (CRC Press 2021).

[7] C. Hill, *Learning scientific programming with Python* (Cambridge University Press 2020).

[8] D. J. Pine, *Introduction to Python for science and engineering* (CRC Press 2019).

[9] W. McKinney, *Python for data analysis: data wrangling with Pandas, NumPy, and IPython* ("O'Reilly Media, Inc." 2012).

[10] A. Géron, *Hands-on machine learning with Scikit-Learn, Keras, and TensorFlow: concepts, tools, and techniques to build intelligent systems* (O'Reilly Media 2019).

[11] D. Ascher and M. Lutz, *Learning Python* (O'Reilly 1999).

[12] T. E. Oliphant, *A guide to NumPy*, volume 1 (Trelgol Publishing USA 2006).

[13] F. Alted, I. Vilata, S. Prater, V. Mas, T. Hedley, A. Valentino, and J. Whitaker, Pytables user's guide, Cárabos Coop 2007 (2002).

[14] D. Y. Chen, *Pandas for everyone: Python data analysis* (Addison-Wesley Professional 2017).

[15] M. Wiebe, M. Rocklin, T. Alumbaugh, and A. Terrel, Blaze: building a foundation for array-oriented computing in Python, in *Proceedings of the 13th Python in science conference. Ed. by Stéfan van der Walt and James Bergstra* (2014), pp. 99–102.

[16] D. Phillips, *Python 3 object oriented programming* (Packt Publishing Ltd 2010).

[17] E. Anderson, Z. Bai, C. Bischof, L. S. Blackford, J. Demmel, J. Dongarra, J. Du Croz, A. Greenbaum, S. Hammarling, A. McKenney *et al.*, *LAPACK users' guide* (SIAM 1999).

[18] P. Virtanen, R. Gommers, T. E. Oliphant, M. Haberland, T. Reddy, D. Cournapeau, E. Burovski, P. Peterson, W. Weckesser, J. Bright *et al.*, Scipy 1.0: Fundamental algorithms for scientific computing in Python, Nature methods 17, 261 (2020).

[19] R. Shankar, *Principles of quantum mechanics* (Springer 2012).

[20] L. I. Schiff, *Quantum mechanics* (McGraw Hill 1969).

[21] N. Zettili, *Quantum mechanics: concepts and applications* (Wiley 2009).

[22] T. F. Jordan, *Quantum mechanics in simple matrix form* (Dover Publications 2005).

[23] J. Baggott, *The quantum cookbook: mathematical recipes for the foundations of quantum mechanics* (Oxford University Press 2020).

[24] M. A. Nielsen and I. L. Chuang, *Quantum computation and quantum information* (Cambridge University Press 2010).

[25] R. A. Horn and C. R. Johnson, *Matrix analysis* (Cambridge University Press 2012).

[26] G. Strang, *Linear algebra and its applications.* (Cengage Learning 2005).

[27] P. A. M. Dirac, *Lectures on quantum mechanics* (Dover Publications 2003).

[28] J. M. Ziman, *Elements of advanced quantum theory* (Cambridge University Press 1975).

[29] H. Wendland, *Numerical linear algebra: an introduction* (Cambridge University Press 2017).

[30] S. H. Friedberg, A. J. Insel, and L. E. Spence, *Linear algebra* (Pearson 2002).

[31] X.-D. Zhang, *Matrix analysis and applications* (Cambridge University Press 2017).

[32] N. J. Higham, *Functions of matrices: theory and computation* (SIAM 2008).

[33] M. Malek-Shahmirzadi, A characterization of certain classes of matrix norms, Linear and Multilinear Algebra 13 (1983).

[34] J. H. M. Wedderburn, The absolute value of the product of two matrices, Bulletin of the AMS 31 (1925).

[35] J. B. Conway, *A course in functional analysis* (Springer 2007).

[36] W. Cheney and D. Kincaid, *Linear algebra: theory and applications* (Jones and Bartlett 2011).

[37] L. Pursell and S. Trimble, Gram-schmidt orthogonalization by Gauss elimination, The American Mathematical Monthly 98 (1991).

[38] N. S. Yanofsky and M. A. Mannucci, *Quantum computing for computer scientists* (Cambridge University Press 2008).

[39] D. P. DiVincenzo, Quantum gates and circuits, Proceedings of the Royal Society of London A 454 (1998).

[40] R. P. Feynman, Quantum mechanical computers, Foundations of Physics 16 (1986).

[41] R. P. Feynman, Simulating physics with computers, in *Feynman and computation* (CRC Press 2018), pp. 133–153.

[42] D. E. Deutsch, Quantum computational networks, Proceedings of the Royal Society of London A 425 (1989).

[43] T. Toffoli, Computation and construction universality of reversible cellular automata, Journal of Computer and System Sciences 15 (1977).

[44] E. Fredkin and T. Toffoli, Conservative logic, International Journal of Theoretical Physics 21 (1982).

[45] V. Vedral, *Introduction to quantum information science* (Oxford University Press 2007).

[46] C. H. Bennett, G. Brassard, C. Crépeau, R. Jozsa, A. Peres, and W. K. Wootters, Teleporting an unknown quantum state via dual classical and Einstein-Podolsky-Rosen channels, Physical Review Letters 70 (1993).

[47] C. H. Bennett and S. J. Wiesner, Communication via one and two particle operators on Einstein-Podolsky-Rosen states, Physical Review Letters 69 (1992).

[48] D. Bohm and Y. Aharonov, Discussion of experimental proof for the paradox of Einstein, Rosen, and Podolsky, Physical Review 108 (1957).

[49] D. M. Greenberger, M. A. Horne, and A. Zeilinger, *Bell's theorem, quantum theory, and conceptions of the universe* (Springer 1989).

[50] Z. Du, X. Li, and X. Liu, Bidirectional quantum teleportation with ghz states and epr pairs via entanglement swapping, International Journal of Theoretical Physics 59, 622 (2020).

[51] H. Prakash, A. K. Maurya, and M. K. Mishra, Quantum teleportation within a quantum network, arXiv preprint arXiv:1210.2201 (2012).

[52] K.-N. Zhu, N.-R. Zhou, Y.-Q. Wang, and X.-J. Wen, Semi-quantum key distribution protocols with ghz states, International Journal of Theoretical Physics 57, 3621 (2018).

[53] Q.-D. Xu, H.-Y. Chen, L.-H. Gong, and N.-R. Zhou, Quantum private comparison protocol based on four-particle ghz states., International Journal of Theoretical Physics 59 (2020).

[54] X. Li and D. Zhang, Multiparty quantum determined key distribution protocol using ghz states, in *2010 International Conference on Networking and Digital Society* (IEEE 2010), volume 1, pp. 203–206.

[55] A. Iqbal and D. Abbott, A game theoretical perspective on the quantum probabilities associated with a ghz state, Quantum Information Processing 17, 1 (2018).

[56] S. Oh, Generation of entanglement in finite spin systems via adiabatic quantum computation, The European Physical Journal D 58 (2010).

[57] W. Dür, G. Vidal, and J. I. Cirac, Three qubits can be entangled in two inequivalent ways, Physical Review A 62 (2000).

[58] A. Einstein, B. Podolsky, and N. Rosen, Can quantum-mechanical description of physical reality be considered complete?, Physical Review 47 (1935).

[59] M. S. Ramkarthik, D. Tiwari, and P. Barkataki, Quantum discord and logarithmic negativity in the generalized N-qubit Werner state, International Journal of Theoretical Physics 59 (2020).

[60] C. E. Shannon, A mathematical theory of communication, The Bell System Technical Journal 27 (1948).

[61] C. Thomas and T. Joy, *Elements of information theory* (Wiley–Blackwell 2006).

[62] V. Vedral, The role of relative entropy in quantum information theory, Reviews of Modern Physics 74 (2002).

[63] C. Fuchs and J. van de Graaf, Cryptographic distinguishability measures for quantum-mechanical states, Information Theory, IEEE Transactions on 45 (1999).

[64] R. Jozsa, Fidelity for mixed quantum states, Journal of Modern Optics 41 (1994).

[65] C. A. Fuchs and C. M. Caves, Ensemble-dependent bounds for accessible information in quantum mechanics, Physical Review Letters 73 (1994).

[66] Q. Wang, Y. Tian, W. Li, L. Tian, Y. Wang, and Y. Zheng, High-fidelity quantum teleportation toward cubic phase gates beyond the no-cloning limit, Phys. Rev. A 103, 062421 (2021).

[67] X. Xu, S. C. Benjamin, and X. Yuan, Variational circuit compiler for quantum error correction, Phys. Rev. Applied 15, 034068 (2021).

[68] S. Chen, L. Wang, S.-J. Gu, and Y. Wang, Fidelity and quantum phase transition for the Heisenberg chain with next-nearest-neighbor interaction, Physical Review E 76, 061108 (2007).

[69] Z.-H. Chen, Z. Ma, F.-L. Zhang, and J.-L. Chen, Super fidelity and related metrics, Central European Journal of Physics 9, 1036 (2011).

[70] W. K. Wootters, Statistical distance and Hilbert space, Physical Review D 23 (1981).

[71] A. Uhlmann, The "transition probability" in the state space of a str-algebra, Reports on Mathematical Physics 9 (1976).

[72] D. Bures, An extension of Kakutani's theorem on infinite product measures to the tensor product of semifinite w star algebras, Transactions of the American Mathematical Society 135 (1969).

[73] C. J. Isham, *Lectures on quantum theory mathematical and structural foundations* (Imperial College Press 1995).

[74] C. Cohen-Tannoudji, B. Diu, and F. Laloe, *Quantum mechanics, vol-1* (Wiley VCH 2006).

[75] R. Horodecki, P. Horodecki, M. Horodecki, and K. Horodecki, Quantum entanglement, Reviews of Modern Physics 81 (2009).

[76] E. Schrödinger, Discussion of probability relations between separated systems, Mathematical Proceedings of the Cambridge Philosophical Society 31 (1935).

[77] E. Schrödinger, Die gegenwärtige situation in der quantenmechanik, Naturwissenschaften 23 (1935).

[78] J. A. Wheeler and W. H. Zurek, *Quantum theory and measurement* (Princeton University Press 2014).

[79] B. M. Terhal, M. M. Wolf, and A. C. Doherty, Quantum entanglement: A modern perspective, Physics Today 56 (2003).

[80] S. Sachdev, *Quantum phase transitions* (Cambridge University Press 2011).

[81] J. Keating and F. Mezzadri, Random matrix theory and entanglement in quantum spin chains, Communications in mathematical physics 252, 543 (2004).

[82] T. Roscilde, P. Verrucchi, A. Fubini, S. Haas, and V. Tognetti, Studying quantum spin systems through entanglement estimators, Physical review letters 93, 167203 (2004).

[83] A. Osterloh, L. Amico, G. Falci, and R. Fazio, Scaling of entanglement close to a quantum phase transition, Nature 416 (2002).

[84] T. J. Osborne and M. A. Nielsen, Entanglement in a simple quantum phase transition, Physical Review A 66, 032110 (2002).

[85] A. Datta and G. Vidal, Role of entanglement and correlations in mixed-state quantum computation, Physical Review A 75, 042310 (2007).

[86] M. Van den Nest, Universal quantum computation with little entanglement, Physical review letters 110, 060504 (2013).

[87] R. Jozsa, Entanglement and quantum computation, arXiv preprint quant-ph/9707034 (1997).

[88] A. Peres, Separability criterion for density matrices, Physical Review Letters 77 (1996).

[89] L. P. Hughston, R. Jozsa, and W. K. Wootters, A complete classification of quantum ensembles having a given density matrix, Physics Letters A 183 (1993).

[90] O. Gühne and G. Tóth, Entanglement detection, Physics Reports 474 (2009).

[91] M. B. Plenio and S. S. Virmani, An introduction to entanglement measures, Quantum Information & Computation 7 (2007).

[92] P. Barkataki and M. S. Ramkarthik, A set theoretical approach for the partial tracing operation in quantum mechanics, International Journal of Quantum Information 16 (2018).

[93] D. Bruß and C. Macchiavello, How the first partial transpose was written?, Foundations of Physics 35 (2005).

[94] M. Horodecki, P. Horodecki, and R. Horodecki, Separability of mixed states: necessary and sufficient conditions, Physics Letters A 223 (1996).

[95] R. Augusiak, M. Demianowicz, and P. Horodecki, Universal observable detecting all two-qubit entanglement and determinant-based separability tests, Phys. Rev. A 77, 030301 (2008).

[96] W. K. Wootters, Entanglement of formation of an arbitrary state of two qubits, Physical Review Letters 80 (1998).

[97] W. K. Wootters, Entanglement of formation and concurrence, Quantum Information & Computation 1 (2001).

[98] A. Buchleitner, A. Carvalho, and F. Mintert, Measures and dynamics of entanglement, The European Physical Journal Special Topics 159 (2008).

[99] P. Rungta and C. M. Caves, Concurrence-based entanglement measures for isotropic states, Physical Review A 67 (2003).

[100] A. Wehrl, General properties of entropy, Reviews of Modern Physics 50 (1978).

[101] A. Peres, *Quantum theory: concepts and methods* (Springer Science & Business Media 2006).

[102] G. Vidal, J. I. Latorre, E. Rico, and A. Kitaev, Entanglement in quantum critical phenomena, Physical Review Letters 90, 227902 (2003).

[103] A. Rényi, On measures of entropy and information, Proceedings of the Fourth Berkeley Symposium on Mathematical Statistics and Probability (1960).

[104] F. Franchini, A. Its, and V. Korepin, Renyi entropy of the XY spin chain, Journal of Physics A: Mathematical and Theoretical 41 (2007).

[105] G. M. Bosyk, M. Portesi, and A. Plastino, Collision entropy and optimal uncertainty, Physical Review A 85, 012108 (2012).

[106] B. Espinoza and G. Smith, Min-entropy as a resource, Information and Computation 226, 57 (2013).

[107] G. Vidal and R. F. Werner, Computable measure of entanglement, Physical Review A 65 (2002).

[108] M. B. Plenio, Logarithmic negativity: a full entanglement monotone that is not convex, Physical Review Letters 95 (2005).

[109] G. K. Brennen, An observable measure of entanglement for pure states of multi-qubit systems, Quantum Information & Computation 3 (2003).

[110] D. A. Meyer and N. R. Wallach, Global entanglement in multiparticle systems, Journal of Mathematical Physics 43 (2002).

[111] H. Li and F. D. M. Haldane, Entanglement spectrum as a generalization of entanglement entropy: Identification of topological order in non-abelian fractional quantum hall effect states, Physical Review Letters 101 (2008).

[112] M. S. Ramkarthik, V. R. Chandra, and A. Lakshminarayan, Entanglement signatures for the dimerization transition in the Majumdar-Ghosh model, Physical Review A 87 (2013).

[113] V. Coffman, J. Kundu, and W. K. Wootters, Distributed entanglement, Physical Review A 61 (2000).

[114] M. Horodecki and P. Horodecki, Reduction criterion of separability and limits for a class of distillation protocols, Physical Review A 59 (1999).

[115] D. C. Mattis, *The many-body problem: an encyclopedia of exactly solved models in one dimension* (World Scientific 1993).

[116] K. Joel, D. Kollmar, and L. F. Santos, An introduction to the spectrum, symmetries, and dynamics of spin-1/2 Heisenberg chains, American Journal of Physics 81 (2013).

[117] J. B. Parkinson and D. J. Farnell, *An introduction to quantum spin systems* (Springer 2010).

[118] R. I. Nepomechi, A spin chain primer, International Journal of Modern Physics B 13 (1999).

[119] A. Auerbach, *Interacting electrons and quantum magnetism* (Springer 1998).

[120] P. Fazekas, *Lecture notes on electron correlation and magnetism* (World scientific 1999).

[121] D. C. Mattis, *The theory of magnetism made simple: an introduction to physical concepts and to some useful mathematical methods* (World Scientific 2006).

[122] T. J. G. Apollaro, G. M. A. Almeida, S. Lorenzo, A. Ferraro, and S. Paganelli, Spin chains for two-qubit teleportation, Physical Review A 100 (2019).

[123] J. P. Barjaktarevic, R. H. McKenzie, J. Links, and G. J. Milburn, Measurement-based teleportation along quantum spin chains, Physical Review Letters (2005).

[124] A. Bayat and S. Bose, Entanglement transfer through an antiferromagnetic spin chain, Advances in Mathematical Physics 2010 (2010).

[125] A. A. Belik, S. Uji, T. Terashima, and E. Takayama-Muromachi, Long-range magnetic ordering of quasi-one-dimensional s= 1/2 heisenberg antiferromagnet sr2cu (po4) 2, Journal of Solid State Chemistry 178, 3461 (2005).

[126] T. Chakraborty, H. Singh, D. Chaudhuri, H. S. Jeevan, P. Gegenwart, and C. Mitra, Investigation of thermodynamic properties of cu (nh3) 4so4 · h2o, a heisenberg spin chain compound, Journal of Magnetism and Magnetic Materials 439, 101 (2017).

[127] P. Pfeuty, The one-dimensional Ising model with a transverse field, Annals of Physics 57 (1970).

[128] H. Bethe, Zur theorie der metalle, Zeitschrift für Physik 71 (1931).

[129] S. Katsura, Statistical mechanics of the anisotropic linear Heisenberg model, Physical Review 127 (1962).

[130] W. Heisenberg, Zur theorie des ferromagnetismus, Zeitschrift für Physik 49 (1928).

[131] C. K. Majumdar, Antiferromagnetic model with known ground state, Journal of Physics C: Solid State Physics 3 (1970).

[132] C. K. Majumdar and D. K. Ghosh, On next-nearest-neighbor interaction in linear chain-I, Journal of Mathematical Physics 10 (1969).

[133] C. K. Majumdar and D. K. Ghosh, On next-nearest-neighbor interaction in linear chain- II, Journal of Mathematical Physics 10 (1969).

[134] P. W. Anderson, Antiferromagnetism. theory of superexchange interaction, Physical Review 79 (1950).

[135] T. Moriya, Anisotropic superexchange interaction and weak ferromagnetism, Physical Review 120 (1960).

[136] I. Dzyaloshinsky, A thermodynamic theory of "weak" ferromagnetism of antiferromagnetics, Journal of Physics and Chemistry of Solids 4 (1958).

[137] F. Schütz, P. Kopietz, and M. Kollar, What are spin currents in Heisenberg magnets?, The European Physical Journal B-Condensed Matter and Complex Systems 41 (2004).

[138] W. K. Wootters, Random quantum states, Foundations of Physics 20 (1990).

[139] P. Forrester, *Log-gases and random matrices* (Princeton University Press 2005).

[140] M. L. Mehta, *Random matrices* (Academic Press 2004).

[141] J. Maziero, Fortran code for generating random probability vectors, unitaries, and quantum states, Frontiers in ICT 3 (2016).

[142] K. Życzkowski, K. A. Penson, I. Nechita, and B. Collins, Generating random density matrices, Journal of Mathematical Physics 52 (2011).

[143] B. Collins and I. Nechita, Random matrix techniques in quantum information theory, Journal of Mathematical Physics 57 (2016).

[144] T. A. Brody, J. Flores, J. B. French, P. A. Mello, A. Pandey, and S. S. M. Wong, Random-matrix physics: spectrum and strength fluctuations, Reviews of Modern Physics 53 (1981).

Subject index